Sobotta/Hammersen · Histology

Sobotta/Hammersen

Histology

A Color Atlas of Cytology, Histology and
Microscopic Anatomy

Frithjof Hammersen, M. D.
Professor and Chairman of the Department of Anatomy
Technical University of Munich

Second edition, revised and enlarged with
499 illustrations mostly in color

Urban & Schwarzenberg · Baltimore–Munich 1980

Urban & Schwarzenberg, Inc.
7 East Redwood Street
Baltimore, Maryland 21202
U.S.A.

Urban & Schwarzenberg
Pettenkoferstraße 18
D-8000 München 2
Germany

A translation of Sobotta/Hammersen · Atlas der Histologie des Menschen
Urban & Schwarzenberg, München–Wien–Baltimore 1979

References:
The figures 95, 179, 180, 211, 217, 222, 226, 227, 244, 246, 248, 254–256, 263–265, 269, 276, 288, 289, 297, 298, 307–309, 313 a b c, 324, 344, 346, 352, 354, 365, 366, 372, 383, 421, 422, 432–435, 450–452, 455–459, 464, 465, 480, 481, 486, 491 and 492 were taken from: Johannes Sobotta, Atlas und Lehrbuch der Histologie und Mikroskopischen Anatomie.
The figures 129, 130, 135, 232, 233, 249, 304, 389, 390, 398 and 401 were taken from: Josef Wallraff, Leitfaden der Histologie des Menschen, 8th edition, Urban & Schwarzenberg, München–Berlin–Wien 1972.
The figures 209 and 377 were taken from: Viktor Patzelt, Histologie, 3 rd. edition Urban & Schwarzenberg, Wien 1948.

Library of Congress Cataloging in Publication Data

Hammersen, Frithjof.
    Histology.

    Translation of Atlas der Histologie des Menschen.
    At head of title: Sobotta/Hammersen.
    Many of the drawings were taken from J. Sobotta's
    Atlas und Lehrbuch der Histologie und mikroskopischen Anatomie.
    Includes index.
    1. Histology––Atlases.   2. Anatomy, Human––Atlases. I. Sobotta, Johannes, 1869–1945, Atlas
        Lehrbuch der Histologie und mikroskopischen Anatomie   II. title.

Printed in Germany by Kastner & Callwey, München

ISBN 0-8067-1742-4 Baltimore
ISBN 3-541-71742-4 München

Preface for the Second Edition

It has been only a few years since this book was first published; its favorable reception has compelled the preparation of this 2nd Edition. This has offered the opportunity for a thorough revision, together with the consideration of many constructive proposals, criticisms and suggestions that have come from referees as well as from our student readers. Publisher and author together have agreed that the original concept of the book should not be altered and should remain what it was originally intended to be – an atlas that deliberately avoids any elaborate text. Nevertheless, a brief description of the technical procedures in preparing a histologic specimen has been added as an introduction to this new edition. These technical remarks are supplemented with a few practical hints on microscopic techniques, the consecutive steps necessary for identifying an unknown histologic specimen and spatial conceptions of histologic structures.

The main aim of this revision, however, was to improve and increase the number of illustrations. All of the electron micrographs have been replaced by newer illustrations and several chapters, including those on the nervous tissue and the liver, have had electron micrographs added to them. Further, those illustrations from the 1st Edition that originated from invertebrates or non-mammals have been replaced in an effort to establish the character of this book as an atlas of *human histology*, even though the common organelles of an invertebrate cell differ only minimally, if at all, from those of a mammalian cell. Altogether 79 electron micrographs have been newly incorporated either as replacements or additions.

In addition, due to the kind understanding and courtesy of the publisher, 132 new color micrographs have been prepared to replace older, not fully satisfying illustrations and to add new pictorial information to several chapters. We were also able to consider another suggestion and incorporated plastic semi-thin sections into this atlas. The sections included are of those organs of which biopsies are now routinely examined with the electron microscope; e.g., the liver and kidneys. The color drawings have been retained in all those cases where they provide better information than photomicrographs although several drawings have been replaced and others newly added. Compared to the 420 figures of the 1st Edition, this atlas now contains a total of 499 figures, comprising 527 individual illustrations. In addition, the tables that most readers have found to be useful have been supplemented by four more tables referring to various aspects of differential diagnosis. The index has also been increased considerably by the addition of many new items.

These alterations were made keeping in mind the basic aim of this atlas: to be a *visual guide for practical work with the microscope.* This is moreover, the reason we did not strive for ultimate technical perfection in every case; because we feel that students must be confronted with specimens displaying the technical quality that can be expected in a routine laboratory for light and electron microscopy.

V

Finally, my sincere thanks are due to Mrs. E. Möhring, who has been associated with this book from its very beginning. Her great experience and technical skill, combined with her constant efforts, made the rapid completion of this edition possible. I also wish to thank my former coworker, Dr. H.-J. Appell, for his critical reading of the 1st Edition and the many helpful suggestions coming from that source.

I am also indebted to Prof. P. Böck and Dr. U. Osterkamp, both of the Department of Anatomy of the Technical University in Munich, for stimulating discussions and the generous supply of material for some new illustrations (as indicated in the captions). Furthermore, it is a pleasure to express my appreciation to the publisher, Mr. Michael Urban, who not only met my wishes with great understanding and generosity, but in addition delegated Prof. H. J. Clemens from his staff to oversee the preparation of this 2nd Edition. It was a real pleasure to work with this experienced colleague over the past 18 months and many thanks are due to him for his excellent cooperation.

It is hoped that the improvement of the illustrations, together with the addition of new ones, has improved the 2nd Edition in such a way that it may serve, like Ariadne's thread, as a reliable guide for the student through the initially confusing labyrinth of cytologic and histologic structures. That is the essential aim and main concern of this book.

Munich, August 1979   Frithjof Hammersen

Contents

Fundamentals of Histologic Techniques

In courses of normal and pathologic histology the student is usually confronted with thin stained sections of tissues and organs that are mounted under a cover slip to provide permanent preparations. In order to interpret these tissue slices correctly and critically, e.g., to be aware of artifacts, the student should be familiar with the fundamentals of histologic techniques.

Without the use of special optical equipment such as phase or interference contrast microscopes, living cells and tissues are almost invisible because of the minimal refraction differences existing among the various cellular and tissue constituents. The living tissues and organs are, therefore, subjected to defined procedures to obtain from them thin stained slices which usually exhibit high contrast. These procedures consist mainly of the following consecutive steps:

Fixation

The fixation should serve at least three different functions:
1. As good a preservation of the tissue as possible in order to stabilize its constituents in an almost *in vivo* condition. Since this is only achievable within certain limits because of the high water content of living materials, a perfect chemical fixative does not exist.
2. An increase in the hardness of the tissue to improve the ease with which it can be cut into thin slices.
3. The killing of all bacteria and other infective agents present in the tissue.

Many of our fixatives, of which the most common is a neutral 5% solution of formaldehyde, are strong precipitants of proteins (e.g., picric acid and mercury bichloride); and hence they coagulate the constituents of cells and tissues. This severe denaturation can be avoided by using a 2.5% solution of glutaraldehyde in a defined buffer (pH 7.4), which is administered – if possible – by vascular perfusion of the organ itself or of an entire experimental animal. This is the most commonly used fixation technique for electron microscopy.

Embedding

In order to obtain sufficiently thin sections (ca. $10\,\mu$m for routine light and ca. 50 nm = 500 Å for electron microscopy) from the fixed and thereby already hardened pieces of tissue, specially designed cutting devices are used (microtomes for light and ultrotomes for electron microscopy); and the specimen must be embedded in a solidifying material with which it can then be cut. As these embedding materials are not soluble in water, it must be removed by placing the tissue step by step through a series of graded alcohol or acetone. This dehydration procedure should be carried out very carefully yet thoroughly, so that finally all the water is removed and is replaced by 100% alcohol or acetone. Thereby the tissue not only becomes additionally hardened, but all its components soluble in alcohol or acetone (e.g., lipids) are dissolved and removed (cf. Fig. 42). During this step of the preparation a particularly great number of artifacts, such as shrinkage (cf. Fig. 8) or disruption of the tissue (cf. Fig. 10), may occur. Paraffin wax is still the most common of embedding materials for light microscopy, whereas various types of polymerizing resins like Epon, Araldite and others are used in electron microscopy.

A particularly good preservation of cellular structures can be achieved by placing the freshly just obtained tissue into liquid nitrogen and then cutting these frozen specimens with a special cryomicrotome, thereby avoiding the disadvantages of the process of dehydration such as shrinkage and the dissolution of lipids.

Staining

Paraffin sections prepared as described above are firmly attached to glass slides, and these are transferred into staining solutions to increase the contrast among the various tissue ingredients. Since most of these staining solutions are aqueous in nature, it is necessary to remove the paraffin by placing the sec-

1

Fundamentals of histologic techniques

Schematic representation of the main steps in preparing a stained thin section from a living tissue as a permanent histologic preparation

Obtaining the living tissue

↓

Fixation (e.g., with 5% solution of formaldehyde)

↓

Dehydration (by increasing concentrations of alcohol)

↓

Embedding (e.g., in paraffin wax)

↓

Cutting and attaching the sections to slides

Mounting the section under a cover slip in an appropriate medium, e.g., canada balsam

↑

Clearing the section (by xylol)

↑

Dehydration of the stained section (by increasing concentrations of alcohol

↑

Staining of the sections

↑

Infiltrating the sections with water (by decreasing concentrations of alcohol)

Removing the paraffin from the section
(by a clearing agent, e.g., xylol)

tion into an appropriate solvent (a clearing agent, e.g., xylol). The solvent must be removed by absolute alcohol, and then the section is transferred into a series of graded alcohol until it is finally placed in water. This is the reverse of the process employed in dehydrating the tissue. When exposed to the staining solutions or mixtures of them, the various tissue constituents take on different colors with different intensities, and this property is greatly influenced by the pH of the staining solution. Besides a series of routine stainings (for these see Figs. 1–4), a great number of special procedures have been developed, among which the so-called histo-topo-chemical methods play a pivotal role in modern histology. These methods allow for a clear identification of a great variety of chemically defined substances such as glycogen, enzymes, lipids, mucopolysaccharides and others at the site of their actual location within cells and tissues, thereby providing much better insight into the biological dynamics of cells.

In electron microscopy, ca. 50 nm (500 Å) thick, so-called ultra-thin sections (mean square area: 0.25 mm²) are placed on circular copper grids (diameter: 3 mm) and are then transferred into the electron beam of the microscope by means of a specially designed specimen holder. The electron microscopic image of such a section is studied on a fluorescent screen, and it can be documented photographically.

These very few remarks must be sufficient to help improve the interpretation of histologic specimens. They are not intended to have the completeness of a textbook of histology.

A Few Remarks to Improve the Interpretation and Identification (Differential Diagnosis) of Histologic Sections

In order to achieve a critical estimate of what a histologic section can prove and what it cannot and how the various structures seen in a section may be interpreted correctly, the student should always be aware of a few, very simple, yet often neglected facts:

1. The histologic section can only provide a sort of "snapshot" produced by the fixation from the continuously changing, dynamic processes occurring in living cells and tissues.
2. The majority of all histologic sections represent a very thin slice of a small piece from a comparably large organ, e.g., the human liver. Due to the often inhomogeneous distribution of certain structures or pathologic processes these cannot be found in every section yet this does not prove that they do not exist.
3. The histologic section can only give a bidimensional image of the usually tridimensional cells and tissues, of whose spaciousness a rather simple technique can sometimes give an impression. In sections that are thicker than the third dimension actually measures, this may be visualized by focusing the microscope objective from the surface of the section down to its bottom (see, e.g., Fig. 317). Immediate conclusions as to the true tridimensional configuration of cells, tissues and their ingredients can be drawn from a single section only in a very few exceptional cases and even then caution is

necessary. This may be illustrated with the following simple examples:

a) Circular profiles, irrespective of whether they are of light or electron microscopic dimensions, could represent transverse sections through cylinders, globules, ellipsoids or cones. If all the profiles seen in one section are of identical size then this would point to cross-sectioned cylinders rather than to any other configuration, because it is not very probable that, e.g., all globules or cones would have been cut at the same level of their circumferences.

b) The occurrence of two nuclei within the same cell is no proof at all that this cell is actually binucleated. In most cases this is due to sections through a bent or curved nucleus cf. Fig. 16), and an apparent "hole" within a nucleus is nothing else but a deep cross-sectioned nuclear indentation.

c) The easiest way to improve the ability to think in three dimensions – and this is essential to interpret histologic sections correctly – is to imagine sections cut in variant planes through familiar tridimensional objects, or even actually to perform such sections (cf. Figs. A–D). If you cut a hard-boiled egg transversely at the site of its two poles (cf. Fig. C) or cut its periphery lengthwise, none of these three sections will show the yolk, but this does not prove that it does not exist. Cross and longitudinal sections through a straight or curved tube (Figs. A and B) may display very different appearances, depending on where and how the sections are cut; and a nearly identical situation may be expected if biological tubes like blood vessels or urinary tubules are involved (cf. Figs. 333–335). Finally, an orange, at least comparable to a secretory unit (acinus) of an exocrine gland like the parotid gland, can appear very different if sectioned in different planes (Fig. D). Many more such simple examples can be found, but these few should stand for the rest of them, as a *pars pro toto*, to illustrate how to develop tridimensional images from histologic sections, which are always and only bidimensional.

4. To identify a so far unknown histologic specimen with certainty and to establish a well-founded differential diagnosis, a few basic rules should be followed that guarantee a systematic procedure in every case.

a) A histologic section should always be inspected thoroughly with the naked eye, because certain organs may be definitely identified if cut in a typical plane like the hypophysis in a midsagittal section (Fig. 398) or a cross-sectioned adrenal gland (Fig. 414).

b) Then should follow an inspection with a simple magnifying glass or, if not available, with the *lowest power* objective of the microscope. During this step pay particular attention

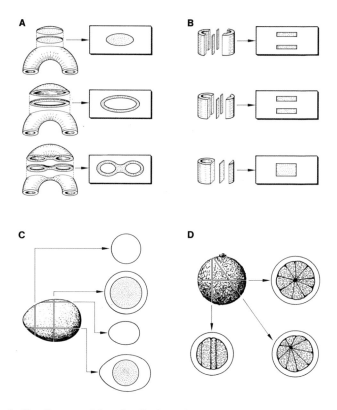

A–D. Cross and longitudinal sections through a bent (A) or straight tube (B), through an egg (C) or an orange (D) can result in section profiles that taken alone do not allow the drawing of any conclusions as to the tridimensional form or the composition of the structure in question (redrawn after Ham and Leeson: *Histology,* 4th ed., London, Pitman Medical Publishing Co. Ltd., 1961).

to the following: Is the section organized into definite layers (corresponding to a "cortex" and a "medulla")? Does a fibrous capsule exist? Can a lumen be found? Does the section contain very different colored areas, or does it show regular folds of its surface?

c) Then the entire section has to be observed with the lowest power objective, i.e., each of its free margins must have been seen. Otherwise one of the most essential questions arising in this context, whether the histologic section displays an epithelium, cannot be definitely decided. If the answer is "Yes" then first of all the epithelium must be carefully identified, because already this guides further considerations in a certain direction, as shown with the following example. If a "simple columnar epithelium" has been found, this section

How to identify an unknown histologic specimen?

could theoretically come from either the digestive canal or from the inner female reproductive organs (uterine tube or uterus). To substantiate the first of these two possibilities, one looks for the characteristic layering of the wall as found in all parts of the digestive tube (cf. Fig. 269). If this can be confirmed, then the section originates with certainty from either the stomach or from the intestinal canal. (For further diagnostic procedures, see Table 13.) In case this typical wall texture is absent, then the section has either been obtained from the gall bladder or the second of the two preliminary diagnoses is true. To decide this (1) search for kinocilia that are characteristic for uterine mucosa and the fallopian tube at defined menstrual stages and (2) relate this feature with the rest of structures, e.g., the occurrence of intensely branched mucosal folds (uterine tube, Fig. 383) or epithelial invaginations as to form simple glands (uterine mucosa, Fig. 387) or the existence of low mucosal folds that anastomose one with another (gall bladder, Fig. 286).

d) These consecutive steps in identifying a histologic specimen can almost always be performed, at least in 95% of all cases, with the lowest power or at most with a medium power objective. If the specimen is not identified with these, it will never be identified at all. This may also be demonstrated with a simple example: To distinguish the various "serous" glands from one another (cf. Table 12, and Figs. 266–268), the pancreas must be included in these considerations. Since the otherwise so reliable "islets" are very rare in the head region and in the uncinate process, or are even entirely lacking in these areas, the inexperienced student thinks the "centro-acinar cells" to be the other most reliable criterion to differentiate the pancreas from the rest of the serous glands. Yet for a beginner it is by far more difficult to identify definitely the centro-acinar cells as such than to evaluate the most simple but much safer criterion, namely the complete absence of salivary (striated) ducts from the exocrine pancreas. But the student searching with the highest power objective for centro-acinar cells must inevitably overlook this characteristic feature due to the very small microscopical field always related to high power objectives, and hence the pancreas will not be correctly identified.

Cytology

Cytology – Stains

Adipose tissue   _Skeletal muscle fiber_

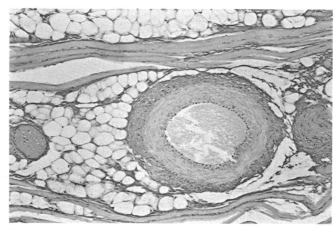

Fig. 1.   _Media of artery_   _Nerve_

Skeletal muscle fiber   _Adipose tissue_

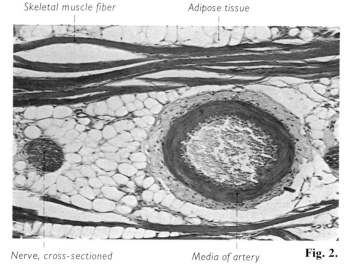

Nerve, cross-sectioned   _Media of artery_   **Fig. 2.**

Fig. 3.

Fig. 4.

Figs. 1–4. Figures 1 to 4 serve as a correlative comparison of the four most common staining procedures on serial sections of the same specimen (cat's tongue muscle). For more details see Table 1. The Mallory-azan preparation shown here is over-stained by the red dye component (due to insufficient differentiation, cf. Fig. 71). In the elastica stain (Fig. 4) the usual counterstaining with nuclear fast red is absent (cf. Fig. 123).

Fig. 1. Hematoxylin and eosin (H and E = H.E.) staining.

Fig. 2. Mallory-azan staining ("azan" named for its two dye components: azocarmine and aniline blue).

Fig. 3. Van Gieson staining is often used in pathology to demonstrate connective tissue proliferations common in various pathological conditions.

Fig. 4. Resorcin-fuchsin staining, like orcein staining, exclusively stains elastic fibers and hence is used for detection of such fibers in the sputum in cases of lung tuberculosis. Magnification of all figures 90×.

Compilation of artifacts due to various technical imperfections. One of the most common is caused by shrinkage during the dehydration process, and therefore largely depends on the different water contents of the tissues.

Fig. 5. Insufficient stretching of the section on the slide by carefully warming the paraffin results in folds that can easily be recognized as such because they appear as darker stained strands that in addition could give a rough estimate of the section's thickness (artery in the capsule of a human vesicula seminalis). The connective tissue in the lower part of the picture shows cracks and clefts. Mallory-azan staining. Magnification 75×.

Fig. 5.

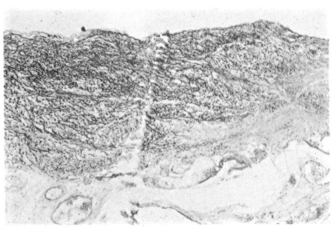

Fig. 6. Scar in a section caused by a nick in the microtome knife (human aortic valve). Resorcin-fuchsin staining. Magnifaction 75×.

Fig. 6.

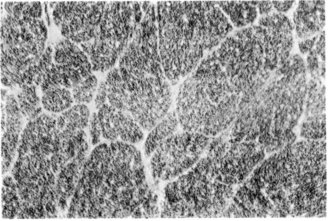

Fig. 7. Irregular thickness due to chattering of the knife (human spinal cord) causes differences in staining intensities, seen here as a lighter staining band traversing the section. Weigert's stain for myelin. Magnification 60×.

Fig. 7.

Cytology – Artifacts

Artificial space caused by shrinkage   Villus

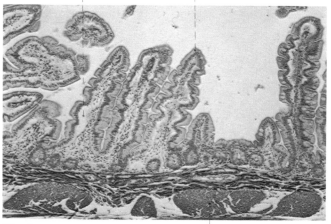

Fig. 8.   Muscularis externa

Compilation of artifacts due to various technical imperfections. One of the most common is caused by shrinkage during the dehydration process, and therefore largely depends on the different water contents of the tissues.

Fig. 8. Extensive clefts caused by shrinkage occurring between the epithelium and the underlying lamina propria of the jejunal villi. Note same artifact between muscle fibers of the tunica muscularis and the adjacent connective tissue (jejunum, man). Mallory-azan staining. Magnification 75×.

Fixative precipitate

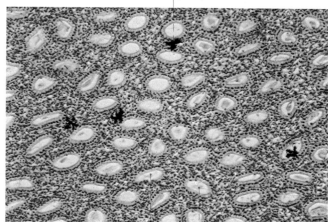

Fig. 9.   Lumen of collecting tubules

Fig. 9. Quite often various fixatives (e.g., formol, sublimate), if not completely removed, can lead to crystalline precipitates seen here as black, irregularly shaped structures (cross section of rabbit renal papilla). H.E. staining. Magnification 75×.

Artificial disruption of muscle fibers

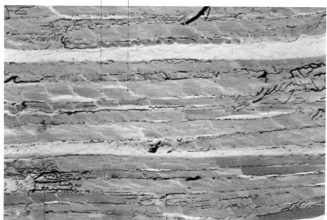

Fig. 10.   Injected blood vessels

Fig. 10. In case of excessive hardening (e.g., by exposing the specimen too long to benzene or benzene-paraffin during the embedding procedure) it becomes friable and the sectioning gives rise to cracks (canine m. rectus femoris, blood vessels injected with ink). H.E. staining. Magnification 75×.

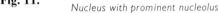
Nucleus of satellite cell

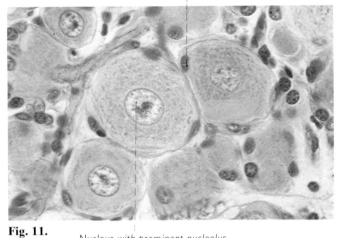

Fig. 11.

Nucleus with prominent nucleolus

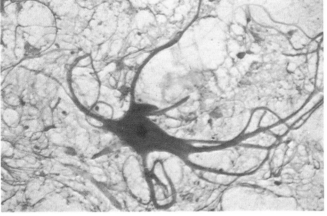

Fig. 12.

Fig. 11. Spinal ganglion cell with its characteristic large and round nucleus with prominent nucleolus swimming in it like an eye. The flat nuclei attached to the ganglion cell surfaces belong to satellite cells, a certain type of peripheral glia cells. This or similar preparations of the primary ovarian follicles are often used to demonstrate the general cell features. Mallory-azan staining. Magnification 600×.

Fig. 12. Multipolar motor neuron of the bovine spinal cord. The cell was isolated by maceration, stained and mounted in its entirety as a squash preparation, thus exhibiting all of its extensions. With common sectioning techniques most of these cell processes would be cut off with only a few lying in the plane of section. Acid fuchsin staining. Magnification 240×.

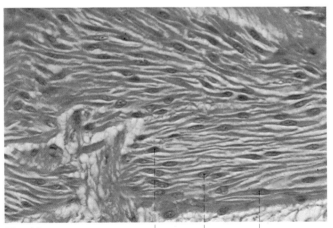

Fig. 13.

Nuclei of spindle-shaped smooth muscle cells

Fig. 13. Longitudinal section of smooth muscle, from rabbit gall bladder, arrayed like a shoal of fish. Note the elliptical nuclei which are often difficult to distinguish from the cytoplasm. Hematoxylin-chromotrop staining. Magnification 380×.

Fig. 14. Unstained spread or "Häutchen" preparation of the isolated eye pigment epithelium (horse) to demonstrate the hexagonal outlines of these cells. The pigment is seen as granules homogeneously distributed in the cytoplasm. Magnification 600×.

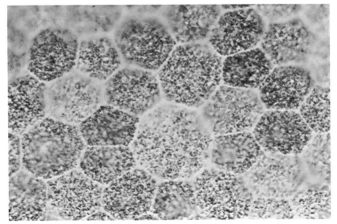

Fig. 14.

Electron microscopy – Schematic representation of a cell

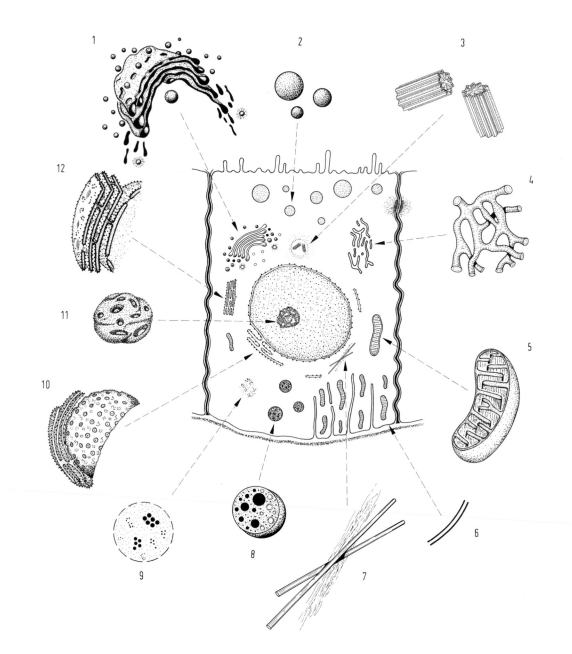

Fig. 15. Schematic representation of an eukaryotic cell with its most common organelles, cytoplasmic constituents and surface specializations (redrawn and extensively modified after Bloom and Fawcett: *Textbook of Histology,* 10th ed., Philadelphia-London-Toronto, Saunders Co., 1975). Because conventional electron micrographs can merely give a bidimensional picture of these structures, they have been depicted in an enlarged tridimensional fashion, neglecting, however, their relative proportions due to the space available. 1 = Golgi apparatus with cisternae, vesicles, vacuoles and coated vesicles; 2 = Secretion granules; 3 = A pair of centrioles (= diplosome); 4 = Smooth endoplasmic reticulum (ER); 5 = Mitrochondrium, crista type; 6 = Cell membrane (plasmalemma); 7 = Microtubules and filaments; 8 = Lysosome; 9 = Glycogen particles and polyribosomes; 10 = Nucleus seen *en face* with its pores and surrounding cisternae of the rough endoplasmic reticulum; 11 = Nucleolus with nucleolonema; 12 = Stacks of cisternae of the rough endoplasmic reticulum (= ergastoplasm). The free surface of the cell bears several irregular microvilli, and the right part of its basal plasmalemma exhibits very regular and deep infoldings.

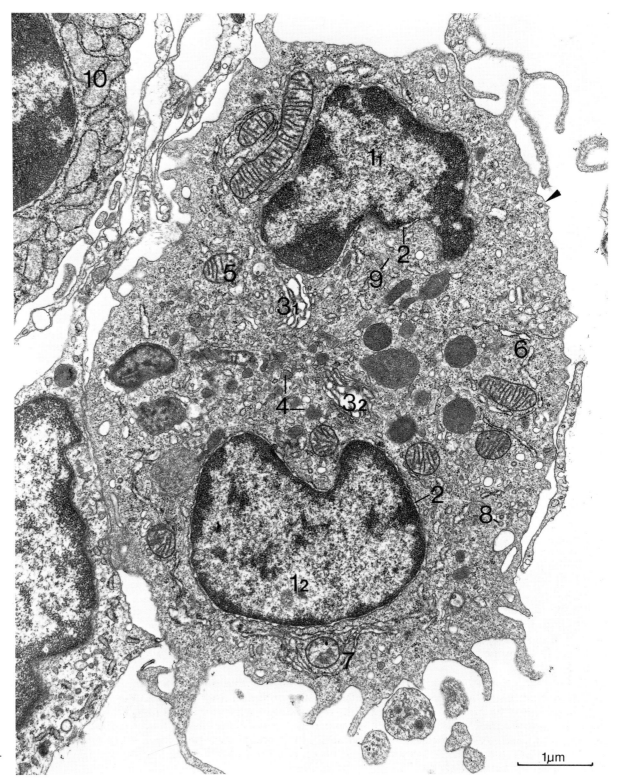

Fig. 16.
Legend s.p. 12.

1µm

Electron microscopy – Cell membrane and its specializations

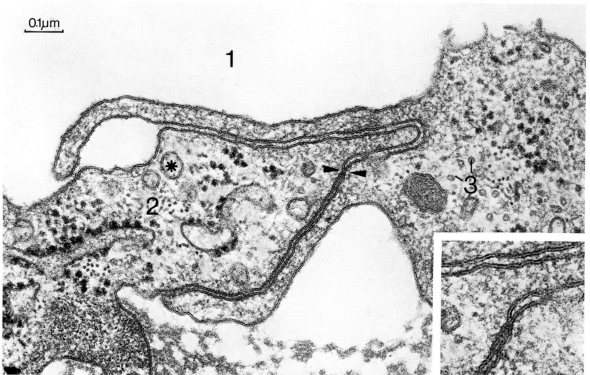

Fig. 17.

Fig. 17. High resolution electron microscopy is necessary to disclose the three-layered appearance of the **cell membrane** (plasmalemma) and its derivatives as, e.g., the micropinocytotic vesicles (*) (endothelial cells of a porcine coronary artery). The endothelial interface clearly shows a point-like fusion of the outer leaflets of the opposing plasmalemmata, a *macula occludens,* a close-up of which is given in the inset. 1 = Vascular lumen; 2 = Filaments, cross-sectioned; 3 = Micotubules, cross-sectioned. Magnification $100,000\times$ and $185,000\times$.

Fig. 18a, b. Microvilli of the intestinal epithelium in longitudinal and cross section from rat jejunum. These finger-like projections are not only very uniform in shape and size (mean length: 0.9μ m, mean diameter: 0.1μ m) but also show a very narrow, yet regular spacing over the entire cell surface. As a whole they form the striated or brush border seen with the light microscope (cf. Fig. 73). As the resolution of this micrograph is rather high, both the trilaminar structure of the cell membrane and the fine parallel filaments positioned in the core of the microvilli are clearly visible. These filaments are possibly contractile and they merge in a perpendicularly oriented filamentous texture, the terminal web, that itself is spread parallel and near to the cell surface. Magnification $78,000\times$ and $72,000\times$.

Fig. 19. Particularly well-developed **basal labyrinth** in an epithelial cell from the proximal convoluted renal tubule (rat). The deep plasmalemmal infoldings form a complex system of irregular intercellular clefts bordering on long but narrow cytoplasmic lamellae that contain numerous, often elongated slender mitochondria (1). Opposite the basal lamina, the cytoplasmic epithelial septa show ill-defined condensations (▶) that correspond to poorly differentiated half-desmosomes (cf. Fig. 50d). Magnification $18,000\times$.

◀ **Fig. 16.** Migratory cell (monocyte) from the connective tissue of a mouse to illustrate the **usual cellular components.** The nucleus clearly shows the outer of its two membranes together with the perinuclear cisterna (2), and the nucleus is cut twice in this section (1_1, 1_2) due to its bent form. In the immediate vicinity of two Golgi complexes (3_1, 3_2) primary lysosomes (4) can be seen, while mitochondria (5) and small cisternae of both the smooth (6) and rough (7) endoplasmic reticulum are scattered throughout the rest of the cytoplasm. Coated vesicles originating from the Golgi apparatus can be seen at ▶ fusing with the cell membrane, thereby contributing to its permanent renewal. 9 = Intracytoplasmic filaments; 10 = Well-developed rough endoplasmic reticulum in an adjacent plasma cell. Magnification $20,000\times$.

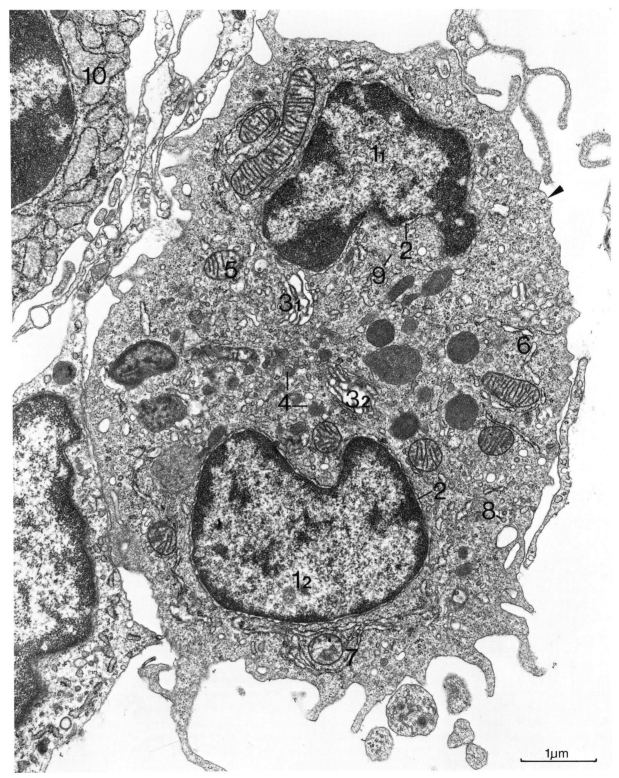

Fig. 16.
Legend s.p. 12.

1μm

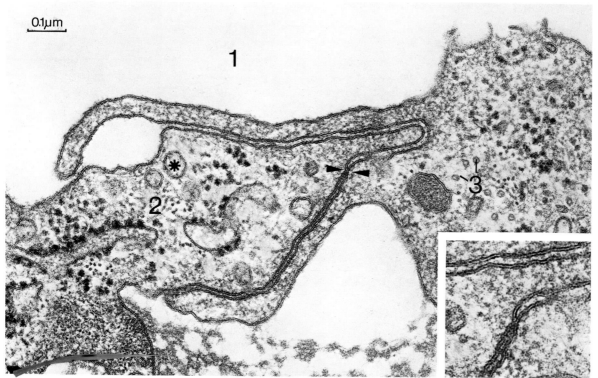

0.1μm

Fig. 17.

Fig. 17. High resolution electron microscopy is necessary to disclose the three-layered appearance of the **cell membrane** (plasmalemma) and its derivatives as, e.g., the micropinocytotic vesicles (*) (endothelial cells of a porcine coronary artery). The endothelial interface clearly shows a point-like fusion of the outer leaflets of the opposing plasmalemmata, a *macula occludens*, a close-up of which is given in the inset. 1 = Vascular lumen; 2 = Filaments, cross-sectioned; 3 = Micotubules, cross-sectioned. Magnification $100,000\times$ and $185,000\times$.

Fig. 18a, b. Microvilli of the intestinal epithelium in longitudinal and cross section from rat jejunum. These finger-like projections are not only very uniform in shape and size (mean length: 0.9μ m, mean diameter: 0.1μ m) but also show a very narrow, yet regular spacing over the entire cell surface. As a whole they form the striated or brush border seen with the light microscope (cf. Fig. 73). As the resolution of this micrograph is rather high, both the trilaminar structure of the cell membrane and the fine parallel filaments positioned in the core of the microvilli are clearly visible. These filaments are possibly contractile and they merge in a perpendicularly oriented filamentous texture, the terminal web, that itself is spread parallel and near to the cell surface. Magnification $78,000\times$ and $72,000\times$.

Fig. 19. Particularly well-developed **basal labyrinth** in an epithelial cell from the proximal convoluted renal tubule (rat). The deep plasmalemmal infoldings form a complex system of irregular intercellular clefts bordering on long but narrow cytoplasmic lamellae that contain numerous, often elongated slender mitochondria (1). Opposite the basal lamina, the cytoplasmic epithelial septa show ill-defined condensations (▶) that correspond to poorly differentiated half-desmosomes (cf. Fig. 50d). Magnification $18,000\times$.

◀ **Fig. 16.** Migratory cell (monocyte) from the connective tissue of a mouse to illustrate the **usual cellular components.** The nucleus clearly shows the outer of its two membranes together with the perinuclear cisterna (2), and the nucleus is cut twice in this section (1_1, 1_2) due to its bent form. In the immediate vicinity of two Golgi complexes (3_1, 3_2) primary lysosomes (4) can be seen, while mitochondria (5) and small cisternae of both the smooth (6) and rough (7) endoplasmic reticulum are scattered throughout the rest of the cytoplasm. Coated vesicles originating from the Golgi apparatus can be seen at ▶ fusing with the cell membrane, thereby contributing to its permanent renewal. 9 = Intracytoplasmic filaments; 10 = Well-developed rough endoplasmic reticulum in an adjacent plasma cell. Magnification $20,000\times$.

Fig. 18.

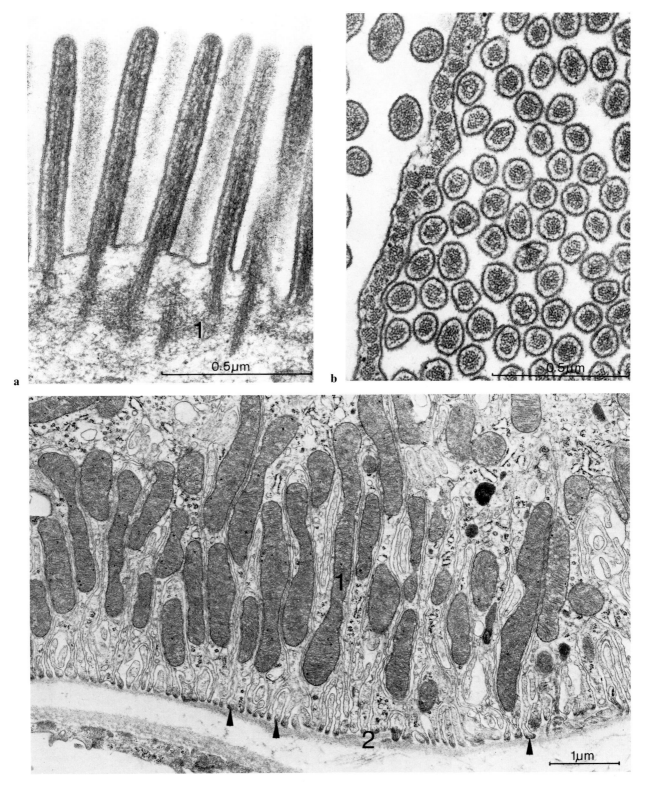

Fig. 19.

Light microscopy – Ergastoplasm and Golgi apparatus

Ergastoplasm

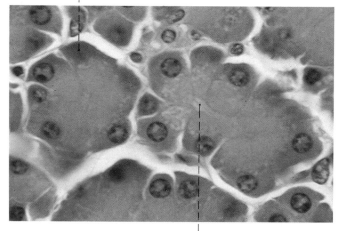

Lumen of serous acinus

Fig. 20. Acini of a canine exocrine pancreas showing extensive basophilia in the abluminal portions of their secretory cells. This staining property is due to the high amounts of RNA occuring in the ribosomes of the **ergastoplasm** (cf. Fig. 26). H.E. staining. Magnification 960 ×.

Nucleus of ganglion cell with prominent nucleolus

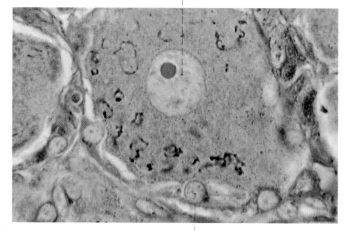

Nucleus of satellite cell

Fig. 21. Neurons of cat spinal ganglion with **Golgi apparatus** appearing as ribbon-shaped deposits. Kolatschev's osmium technique with nuclei and nucleoli counterstained by safranin. Magnification 960 ×.

▶

Fig. 22. Parts of an acinus of a rat exocrine pancreas. In this low-power electron micrograph the **ergastoplasm** (1) appears as a system of electron-dense membranes (due to the ribosomes attached to their surface) that enclose more or less parallel, slightly curved spaces of variant width. 2 = Nucleus; 3 = Secretion granules. Magnification 12,500 ×.

Fig. 23. Low-power electron micrograph of two **Golgi complexes** (1) in an epithelial cell from an intercalated duct of the feline submandibular gland. Each of these complexes corresponds approximately to one of the hook- or ribbon-shaped structures seen with the light microscope after osmication as shown in Fig. 21. 2 = Nucleus; 3 = Mitochondria; 4 = Intercellular space with desmosomes (▶). Magnification 32,000 ×.

14

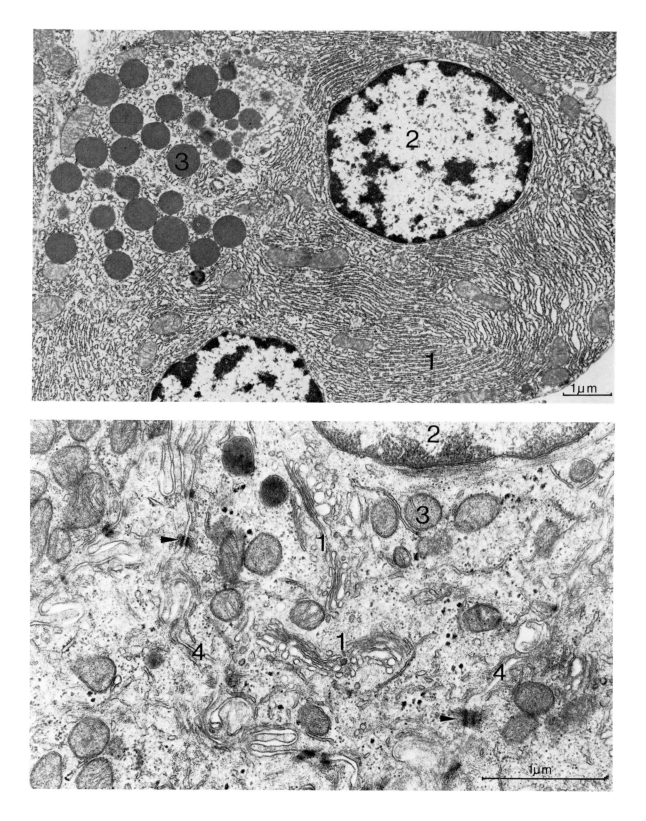

Fig. 22.

Fig. 23.

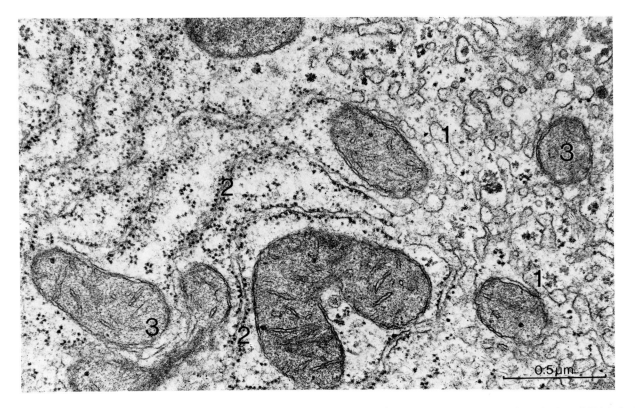

Fig. 24. Part of a liver cell (hepatocyte) from the rat that contains *"rough" or* **granular endoplasmic reticulum** *(ER)* in its left half and *"smooth" or* **agranular ER** in its right half. The smooth ER is not studded with ribosomes, and it appears here in the form of elongated oval membrane profiles (1). 2 = "Rough" ER; 3 = Mitochondrium. Magnification 53,000×.

Fig. 25. Well-developed **smooth endoplasmic reticulum** in the apical portion of an epithelial cell from a murine trachea, disclosing its tube-shaped elements at →. 1 = Mitochondrium; 2 = Microperoxisome. Magnification 44,000×.

Fig. 26. Ergastoplasm in an exocrine pancreatic cell of the rat. The ergastoplasm consists of strictly parallel membranes intensely studded with ribosomes that enclose narrow elongated cavities, the cisternae, which occasionally show vacuolar dilatations at their free edges (*). A rough ER as well developed as it is represented by the ergastoplasm is always the structural equivalent for a high rate of protein synthesis. The products of this activity, however, are not needed for the cell metabolism itself; but they serve as "proteins for export" for the formation of secretion products, intercellular substances, etc. 1 = Nucleus. Magnification 38,000×.

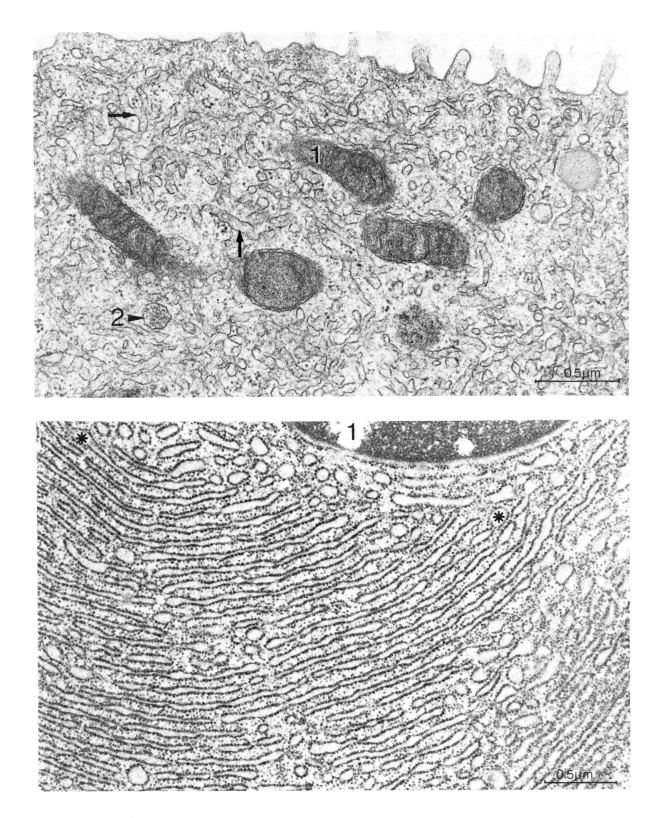

Fig. 25.

Fig. 26.

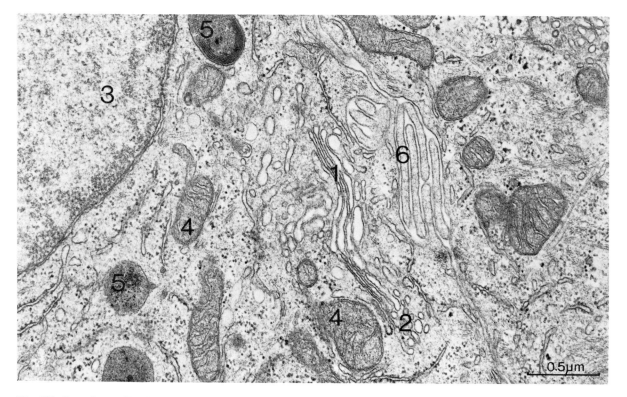

Fig. 27. Prominent **Golgi complex** in an epithelial cell from a salivary duct of the feline submandibular gland. This organelle consists of a low stack (dictyosome) of flattened saccules or cisternae (1), together with the surrounding vesicles and vacuoles (2) that are formed by budding off from the free edges of the Golgi cisternae (cf. Fig. 15). 3 = Nucleus; 4 = Mitochondria; 5 = Lysosomes; 6 = Interdigitating processes of adjoining cells. Magnification 37,000×.

Fig. 28. Several **Golgi complexes** in the apical cytoplasm of an intestinal epithelial cell (rat), whose cisternae (1) display marginal dilatations (▶) filled with an electron-dense material. Note the vesicles [primary lysosomes (2)] with identical contents in close vicinity. Magnification 29,500×.

Fig. 29. a) Several **Golgi complexes** (2) positioned very close to the nucleus (1) in a muscular satellite cell from the feline tongue. Magnification 42,000×.
b) Golgi apparatus showing prominent stack of cisternae together with closely associated vesicles at their free edges. This topographical situation already suggests that the vesicles are being formed by budding off from the Golgi cisternae (blood cell of the piscine leech, *Piscicola geometra*). Magnification 60,000×.

18

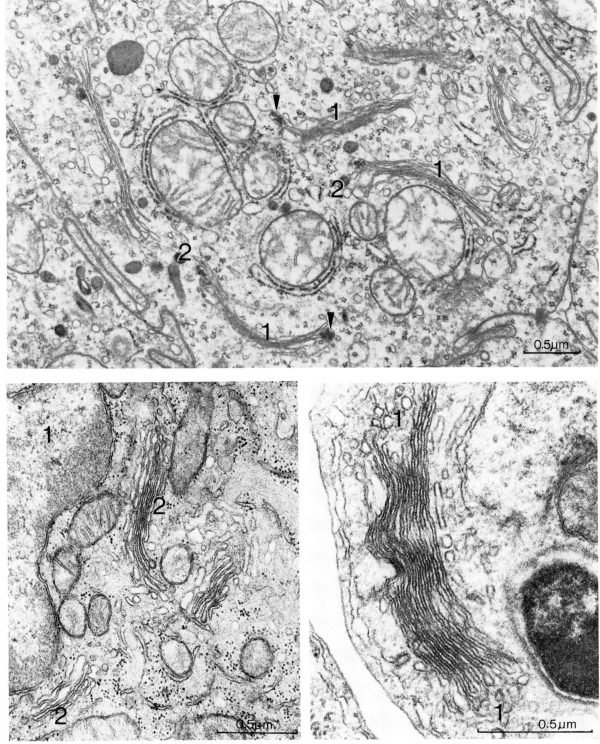

Fig. 28.

Fig. 29.   a   b

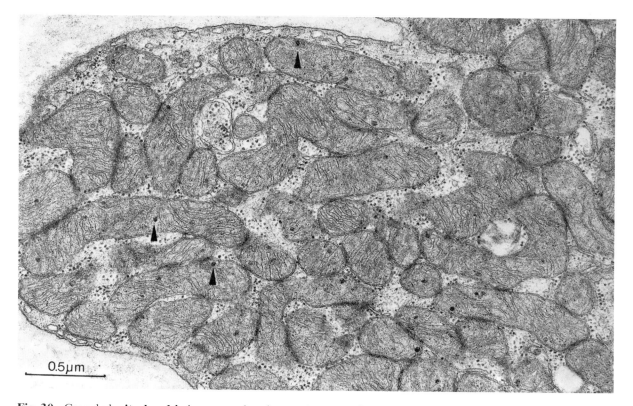

Fig. 30. Crowded **mitochondria** in a sarcoplasmic protuberance of a skeletal muscle fiber from a feline tongue. The differences in shape and size of these organelles are partially realistic and partially caused by the plane of the section. Several of these mitochondria contain dense granules (▶) within their matrix, which is subdivided by numerous membrane-bound electron lucent narrow cavities, the cristae intramitochondriales. The bilaminar nature of the mitochondrial membrane is only occasionally visible (cf. Fig. 48b) Magnification 41,000×.

Fig. 31. Mitochondria from different types of cells to illustrate fine structural details and the differences in shape and arrangement of their internal structure.
a) Liver mitochondria (rat) with a poorly developed **internal structure** consisting of irregularly arranged cristae. This is a characteristic feature of hepatic mitochondria in several species. The cristae represent disc-like folds or pleats of the inner mitochondrial membrane projecting into the matrix. Magnification 63,000×.
b) Well-developed internal structure in epithelial mitochondria. At ▶ the site of an invagination of the inner mitochondrial membrane is clearly visible. Magnification 53,000×.
c) Mitochondria in an endocrine cell of the zona fasciculata (suprarenal gland, rat) exhibiting an internal structure mainly consisting of vesicles. Hence these mitochondria are classified as **"vesicular type"**. 1 = Nucleus; 2 = Golgi apparatus; 3 = Lipid droplets. Magnification 20,000×.
d) Mitochondria classified as **"tubular"** because their internal structure consists of tubules that are often twisted and curved (→). They usually appear, therefore, as roundish or ovoid membrane profiles (suprarenal gland, cat). 1 = Nucleus; 2 = Lipid vacuole. Magnification 39,500×.

Fig. 31.

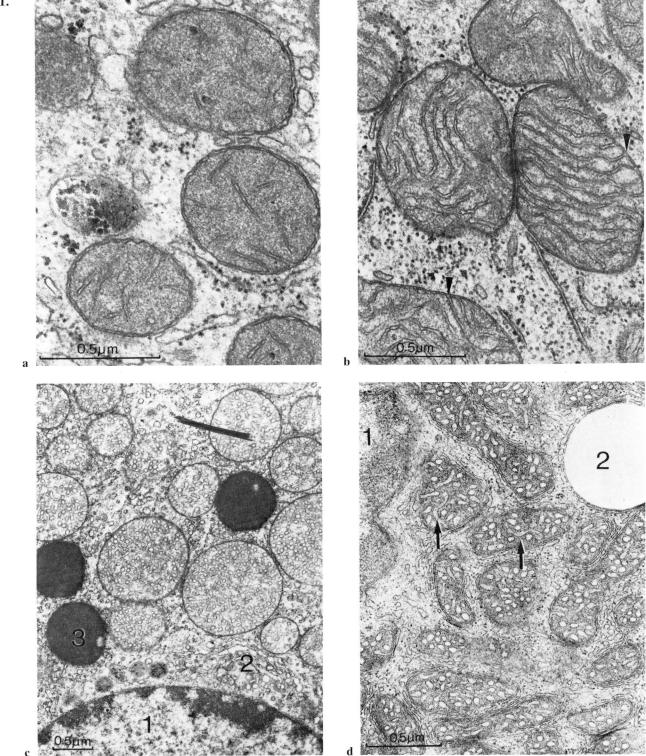

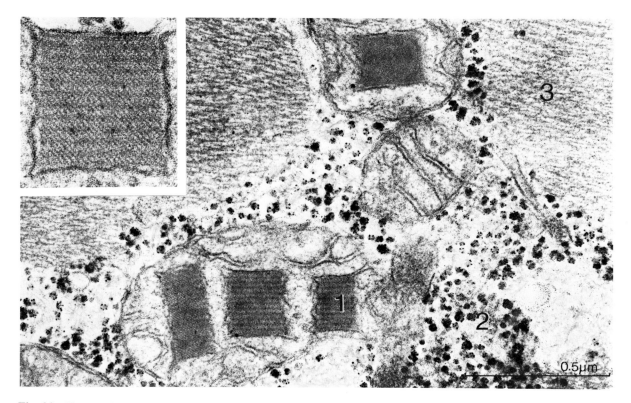

Fig. 32. Human skeletal muscle mitochondria whose cristae contain very regularly outlined electron-dense inclusions (1). At higher magnifications (inset) these rectangular prismatic corpuscles disclose a paracrystalline substructure. 2 = Glycogen particles; 3 = Tangential section of a myofibril. Magnification 75,000× and 183,000×.

Fig. 33. **Secondary lysosomes** (1) in an epithelial cell from a salivary duct of a feline submandibular gland. Though lysosomes can definitely be identified as such only by means of cytochemical techniques as, e.g., the demonstration of acid phosphatase, the pleomorphic contents of these membrane-bound granules allows for their classification as secondary lysosomes. 2 = Intracytoplasmic filaments; 3 = Desmosomes at the intercellular space; 4 = Nucleus. Magnification 26,500×.

Fig. 34. a) **Autophagic vacuole** in a rat hepatic cell, which contains profiles of the smooth ER (1) together with an intact mitochondrium (2). After fusion with primary lysosomes, such vacuoles are called "cytolysosomes" or "autolysosomes." Magnification 48,000×.
b) Rather large **secondary lysosomes** surrounded by an ill-defined membrane. Their globular contents (lipids) are also found lying free within the cytoplasm (1). These structures probably represent cytolysosomes converting into **residual bodies** (reticulum cell from rat thymus). Magnification 25,000×.

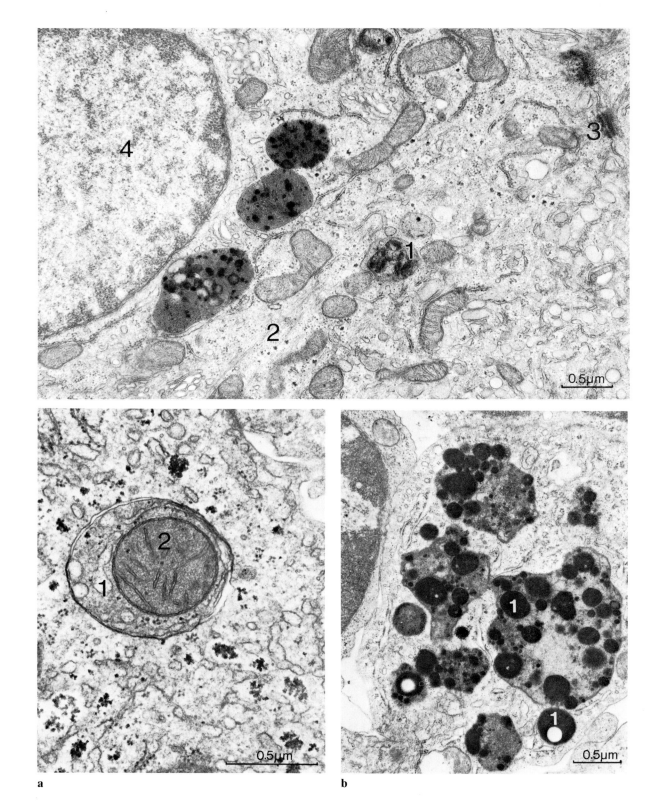

Fig. 33.

Fig. 34.

a

b

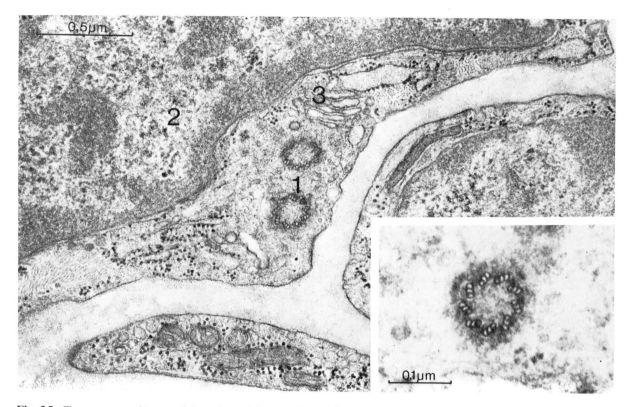

Fig. 35. Transverse and tangential sections of the **centrioles** (1) constituting the diplosome in a fibroblast from feline connective tissue. As in most cases, neither of the two centrioles is cut exactly parallel or perpendicular to the long axis of the tubular units forming the wall of this organelle. Note that in this case, contrary to the rule, the long axes of the centrioles are not perpendicular. 2 = Nucleus; 3 = Part of a dictyosome. The inset shows a centriole exactly cross-sectioned, thereby disclosing that its "wall" is composed of nine longitudinally oriented units (= triplets), each of which consists of three microtubules joined together. Magnification 50,000 × and 110,000 ×.

►

Fig. 36. Microtubules from a mitotic spindle in longitudinal section (thymocyte, rat). The wall of these tubules (inner diameter: 6–10 nm, outer diameter: 20–26 nm) is not composed of a unit membrane but instead consists of 11–13 parallel filaments, each of which is made up of protein molecules. 1 = Part of a chromosome. Magnification 62,000 ×.

Fig. 37. a) Small bundles (1) of fine parallel **filaments** (diameter, 6 nm) in an intestinal epithelial cell of the rat. Magnification 56,000 ×.
b) Structurally similar filaments arranged into much coarser bundles (1) that correspond to the **tonofibrils** of light microscopy are found in human epidermal cells (cf. Fig. 427). 2 = Nucleus; 3 = Melanin granules. Magnification 32,000 ×.

Fig. 36.

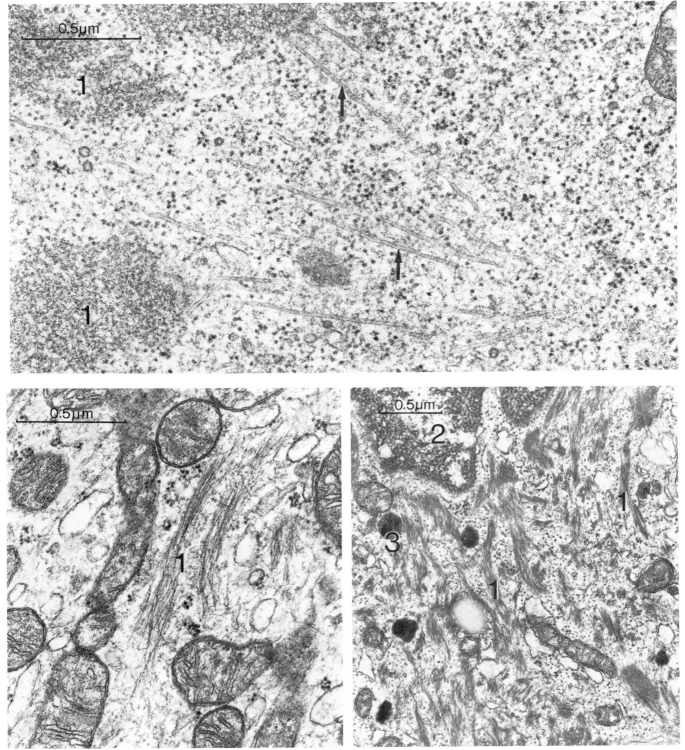

Fig. 37a.

Fig. 37b.

Cytology – Cell inclusions

Rod-shaped crystalloids in interstitial cells

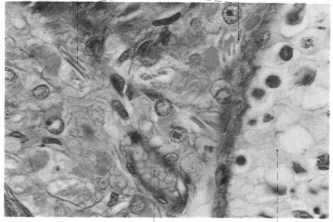

Fig. 38.   *Venule filled with erythrocytes*   *Seminiferous tubule*

Nucleus of a ganglion cell

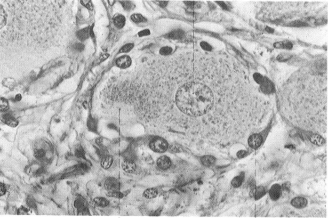

Fig. 39.   *Axon hillock*   *Nucleus of a satellite cell*

Nucleus of hepatic cell

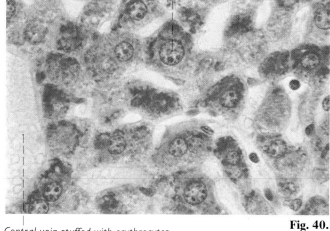

Central vein stuffed with erythrocytes   **Fig. 40.**

Secretion granules

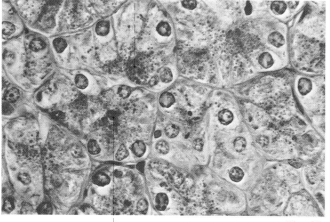

Lumen of serous acinus   **Fig. 41.**

Fig. 38. Section of seminiferous tubule from human testis with adjacent interstitial cells (Leydig cells) displaying characteristic rodshaped proteinaceous crystalloids known as the **crystals of Reinke.** Mallory-azan staining. Magnification 600×.

Fig. 39. Human spinal ganglion cell with **lipofuscin granules** accumulated above the axon hillock (see also Fig. 169). This endogenic pigment is related to the lysosomal population originating from residual bodies and was previously known as "detrition" pigment. Mallory-azan staining. Magnification 600×.

Fig. 40. Ratliver cells selectively stained for **glycogen,** which appears here as fine granules or in the form of coarser clumps colored magenta (courtesy of Prof. H. J. Clemens, Munich). PAS-hemalum staining. Magnification 600×.

Fig. 41. Serous alveoli of human submandibular gland with many red-stained **secretory granules** of various sizes showing different degrees of acidophilia. Mallory-azan staining. Magnification 600×.

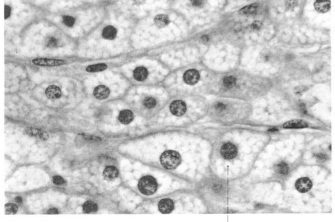

Fig. 42.   *Lipid vacuole*

Epithelial cells with melanin granules   *Connective tissue*

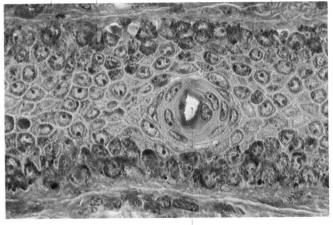

Fig. 43.   *Duct of a sweat gland*

Alveolar phagocytes filled with hemosiderin granules

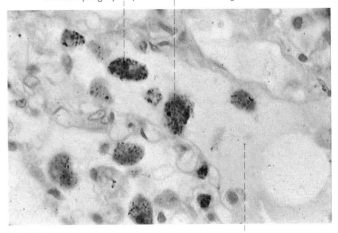

Fig. 44.   *Lumen of alveolus*

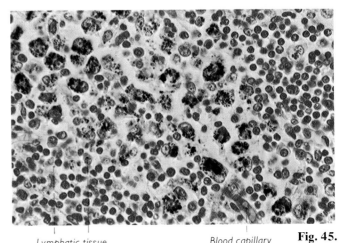

Lymphatic tissue   *Blood capillary*   **Fig. 45.**

Fig. 42. Cluster of polygonal cells crowded with translucent **vacuoles,** hence referred to as "spongiocytes" (human suprarenal zona fasciculata). The vacuoles are due to the extraction of lipids by the organic solvents used in routine histology techniques (cf. p. 1). Mallory-azan staining. Magnification 600×.

Fig. 43. Section parallel to the boundary between the epithelium and connective tissue of the skin (rhesus monkey). The basal epithelial cells contain brown-black pigment granules **(melanin).** Compare also with the heavily pigmented cells of the hair cortex (Fig. 433). Mallory-azan staining. Magnification 600×.

Fig. 44. Human lung alveoli showing siderophages or heart failure cells that develop in chronic pulmonary congestion by a progressive incorporation of the hematogenous pigment **hemosiderin.** H.E. staining. The iron is demonstrated by the Turnbull blue reaction, an important technique in pathological histology. Magnification 600×.

Fig. 45. Section of a medullary sinus of a lymph node (human lung) containing numerous macrophages filled with dust particles, an **exogenous pigment.** This phenomenon is known as anthracosis (anthrax = coal). Mallory-azan staining. Magnification 380×.

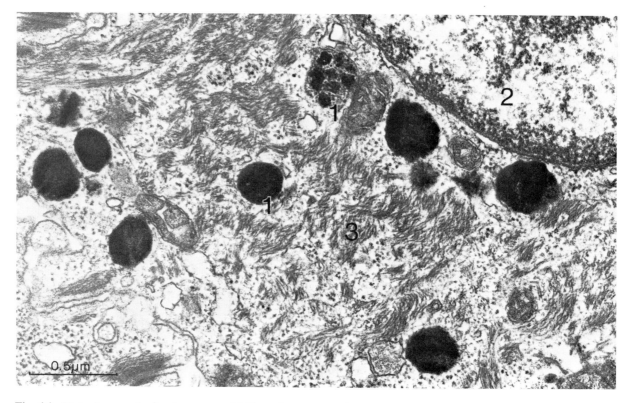

0.5µm

Fig. 46. **Melanin granules** [melanosomes (1)] in various stages of maturation in a human epidermal cell. 2 = Nucleus; 3 = Intracytoplasmic filaments. Magnification 45,500 ×.

▶

Fig. 47. Membrane-bound **secretion granules** (1) of various sizes in the apical portions of rat exocrine pancreatic cells. 2 = Nucleus; 3 = Lumen of acinus; 4 = Mitochondria. Magnification 18,000 ×.

Fig. 48. a) **Lipid droplets** of various sizes whose limiting membranes are poorly defined due to the high yet identical electron density of the membranes and the contents they enclose (renal tubular cell, rat). Magnification 17,500 ×.
b) **Glycogen particles** occurring individually and not aggregated into rosettes as shown in this micrograph are referred to as β-particles (mean diameter: 15–30 nm). Note the distinct matrix granules (▶) within the mitochondria (from a human skeletal muscle fiber). Magnification 42,000 ×.

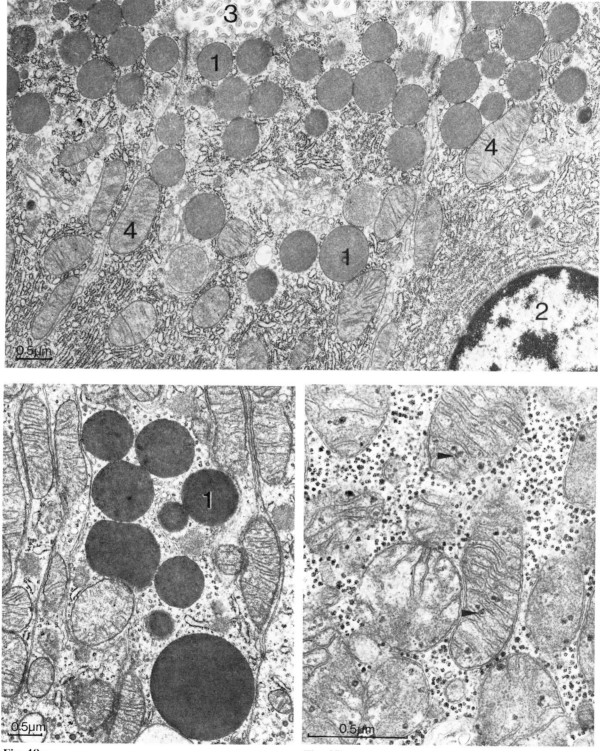

Fig. 47.

Fig. 48a.

Fig. 48b.

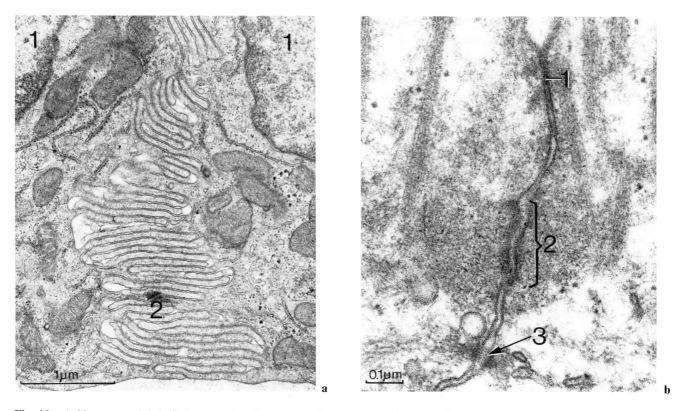

Fig. 49. a) Numerous tightly fitting cytoplasmic processes from an extensive area of **interdigitations** that serve as a means for intercellular attachment between these two epithelial cells (salivary duct from feline submandibular gland). 1 = Nucleus; 2 = Desmosome. Magnification 25,000×.

b) **Junctional complex** between adjoining intestinal epithelial cells (rat). This attachment device corresponds to the terminal bar seen with the light microscope (cf. Fig. 70), and it often consists of three consecutive components (1, 2, 3). As each of these is cut slightly oblique in this section they all appear somewhat blurred. 1 = Zonula occludens; 2 = Zonula adherens; 3 = Macula adherens (= desmosome). Magnification 95,000×.

Fig. 50. a) Spot-like discrete fusion (▶) of the outer leaflets of the opposing cell membranes, so-called **macula occludens,** at an interendothelial cleft (myocardial capillary, cat). 1 = Capillary lumen; 2 = Myocardial mitochondrium. Magnification 130,000×.

b) Endothelial interface from rabbit thoracic aorta displaying fusion of the outer leaflets of the adjoining cell membranes over a longer distance (1), so-called **zonula occludens** (= tight junction). Magnification 265,000×.

c) Characteristic **desmosome** between intestinal epithelial cells (rat). The trilaminar structure of the cell membranes is clearly visible over the entire length of the desmosome, and the intercellular space shows a central dense line possibly corresponding to the intercellular cement. Subjacent to the cytoplasmic surface of the desmosomal plasmalemmata is found an electron-dense plaque (1) consisting of a fine feltwork of filaments into which numerous tonofilaments (2) seem to merge and to terminate therein. Magnification 120,000×.

d) Basal cytoplasmic projection of an epidermal cell from human stratum germinativum. Its cell membrane bears several spot-like condensations (1), so-called half- or **hemidesmosomes,** upon which bundles of tonofilaments converge. 2 = Basal lamina. Magnification 48,000×.

Fig. 50a.

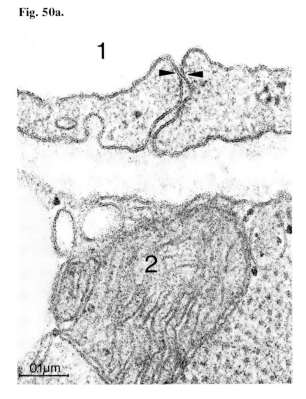

Fig. 50b.

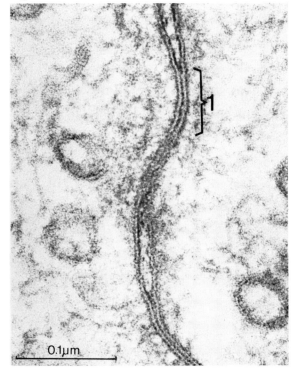

Fig. 50c.

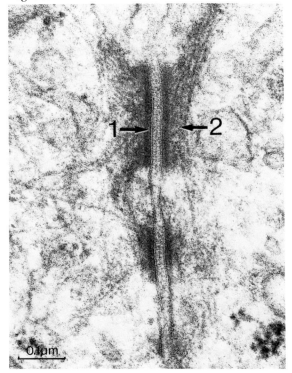

Fig. 50d.

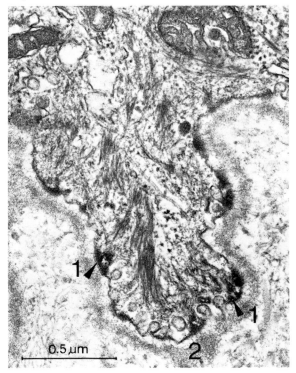

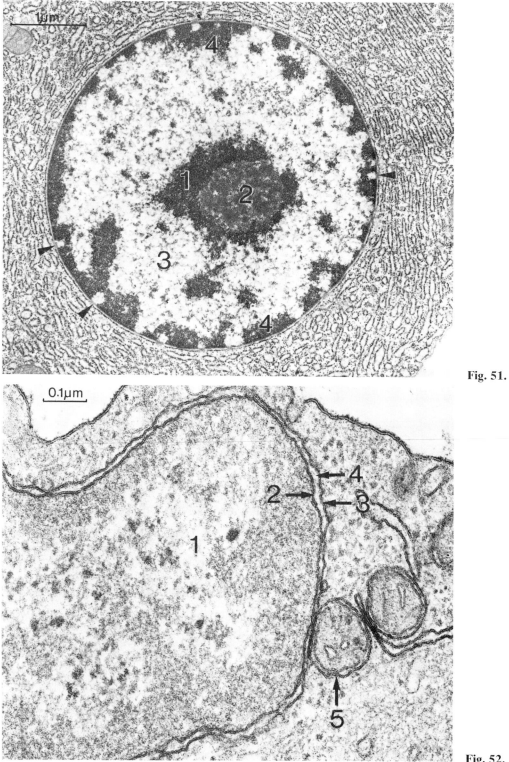

Fig. 51.

Fig. 52.

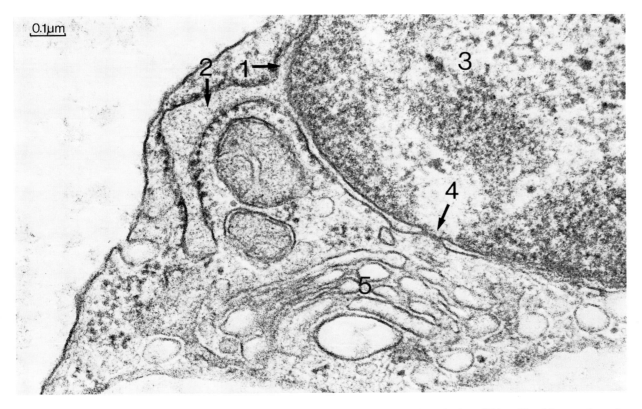

Fig. 53. Prominent communication of the perinuclear space (1) with the rough endoplasmic reticulum (2) in a fibroblast nucleus. 3 = Nuclear matrix; 4 = Nuclear pore; 5 = Golgi cisternae. Magnification 90,000×.

◄

Fig. 51. Typical vesicular **nucleus** with a **nucleolus** (2) surrounded by **chromatin** (1), with an electron lucent nuclear matrix [karyoplasm (3)], and chromatin particles [karyosomes (4)] subjacent to the nuclear envelope that exhibit distinct interruptions (►) at the site of the nuclear pores (exocrine pancreatic cell, rat). While the outer membrane of the nuclear envelope is clearly demarcated from its surroundings due to its being richly equipped with ribosomes, the inner membrane is only poorly defined due to the closely attached chromatin. (For more details of the nuclear envelope, see the following micrographs.) Magnification 19,500×.

Fig. 52. Part of an endothelial nucleus (1) clearly exhibiting the trilaminar nature of each of its two membranes, i.e., inner and outer (2, 3), that together with the perinuclear space (4) constitute the **nuclear envelope.** Note that the three-layered appearance of the inner and outer mitochondrial membranes (5) is also apparent. Magnification 110,000×.

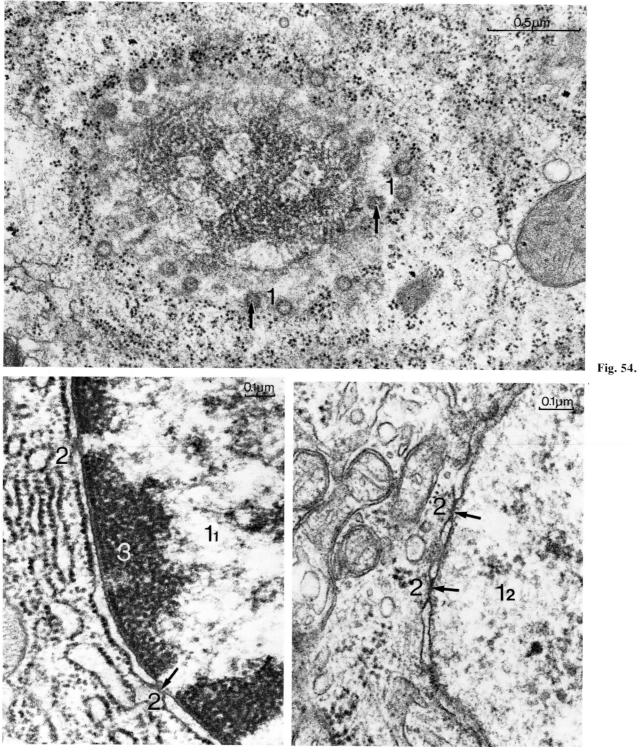

Fig. 54.

Fig. 55a.

Fig. 55b.

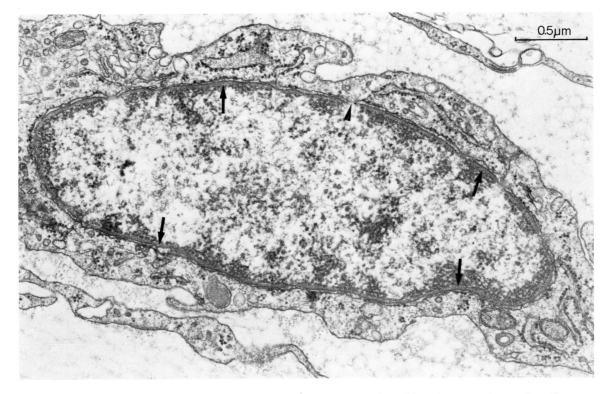

Fig. 56. Prominent **fibrous lamina** (→) in a fibroblast nucleus (subcutaneous tissue, rat) positioned as a continuous layer between the inner nuclear membrane and the chromatic substance. Only at the sites of the nuclear pores (▶) does the lamina fibrosa reveal corresponding circular disruptions. Magnification 38,000×.

◄

Fig. 54. *En face* view of a nucleus from a liver cell (rat) exhibiting numerous **pores** (1). These appear as circular openings in routine electron micrographs (inner diameter: 35 nm) and show a central knob-like condensation (→) in favorable sections. The pores are potential avenues between nucleo- and cytoplasm and are often referred to as a "pore complex" due to their intricate substructure, which, however, is not resolved in this micrograph. Magnification 48,000×.

Fig. 55a, b. Parts of nuclei from an exocrine pancreatic cell (1_1) and from a ganglion cell (1_2) of the rat, revealing distinct **pores** (2) around whose periphery the inner and outer nuclear membranes are continuous with each other, thereby sealing the perinuclear space. In addition the pores are often bridged by a delicate "membrane," the diaphragm; and immediately adjacent to the inner surface of the nuclear envelope a fine filamentous layer, the lamina fibrosa, can be found in some cell types (Fig. 55a). 3 = Chromatin. Magnification 78,000× and 86,000×.

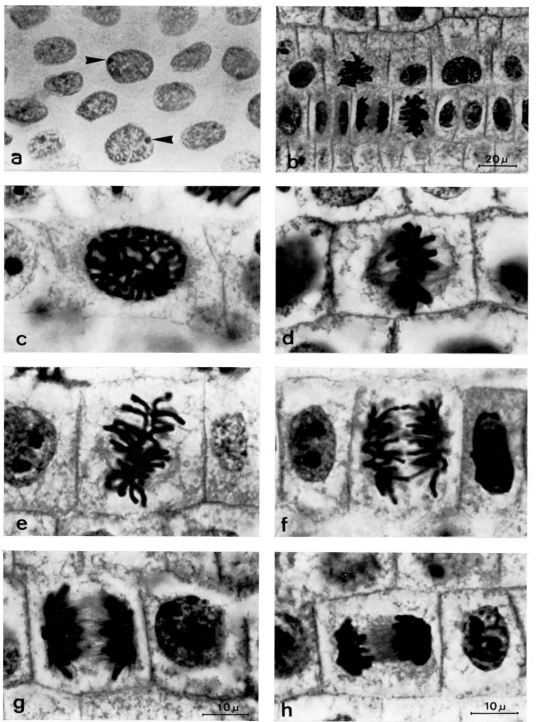

Fig. 57.

Fig. 57. a) Spread or "Häutchen" preparation of a human amnion. The deoxyribonucleic acid (DNA) of the nucleus is stained by the Feulgen reaction and shows a small condensation closely apposed to the nuclear membrane. This represents the **Barr body** that corresponds to one of the two X chromosomes occurring in normal cells of female subjects. Feulgen staining. Magnification 1,250×.
b–h) Different stages of **karyokinesis** from mitotic cell divisions, whose final stage, the cytokinesis, is not shown. The last is the division of the cytoplasm and results in the definite formation of two separate daughter cells. For the histological demonstration of cell divisions, rapidly growing tissues with a high rate of mitoses as cell cultures, embryos or, as in this case, germinating plant seedlings are used.
b) Low-power micrograph from the tip of the root of a bean seedling (Vicia faba) showing many closely apposed cells whose nuclei exhibit different stages of karyokinesis. On either side of the lower row of cells one can see two cells, each of which is exactly half the size of the parent cell and therefore can be assumed to be the daughter cells resulting from a complete mitosis. On the left side these cells lie adjacent to an early telophase (cf. also Fig. 57 h) while they are adjoining a late metaphase on the right side. Iron-hematoxylin staining. Magnification 500×.
c) Nuclear division showing intermediate stage of a **prophase.** At this time the chromosomes are arranged in closely intertwined coiled threads with "neither ends nor beginnings."
d) **Metaphase** in lateral view with the chromosomes situated midway in the spindle and aligned in the equatorial plate. The prominent mitotic spindle consists of fibers that correspond to bundles of microtubules, which connect the kinetochore of each of the chromosomes with the centrioles located at the poles.
e) Late metaphase seen obliquely; hence the precise orientation of the chromosomes cannot be fully recognized. At certain points sister chromatids begin to separate. These originate from each of the chromosomes by reduplication in the S-phase and represent the definite chromosomes of the two future daughter cells.
f) Early **anaphase** with all sister chromatids separated into daughter chromatids that have been pulled apart and moved toward the poles.
g) Late anaphase with prominent continuous spindle fibers connecting the two centrioles. The daughter chromatids that are the definite chromosomes of the reconstructing nuclei of the daughter cells are already losing their individuality.
h) Early **telophase** with an increasing clumping of chromosomes into a homogeneous, intensely staining basophilic mass. The continuous spindle fibers are still clearly visible. Figs. 57 c–h: Iron-hematoxylin staining. Magnification 1,250×.

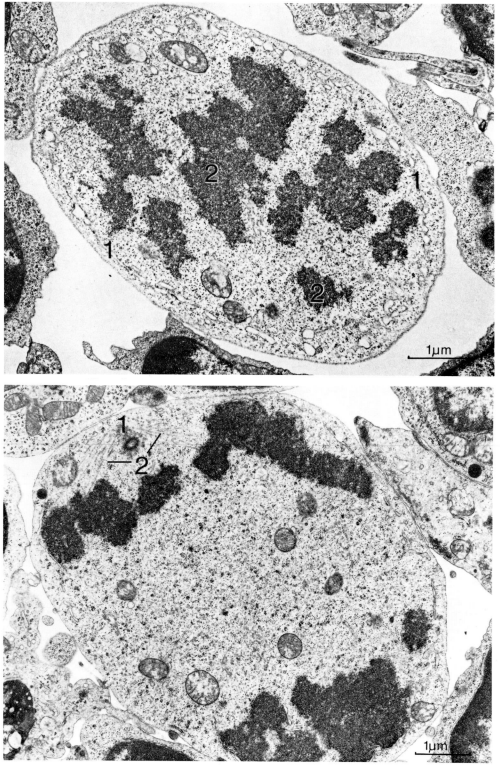

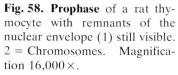

Fig. 58. Prophase of a rat thymocyte with remnants of the nuclear envelope (1) still visible. 2 = Chromosomes. Magnification 16,000×.

Fig. 59. Thymocyte in late **anaphase** (cf. Fig. 57 g) showing one of its centrioles (1) in a polar position together with parts of the mitotic spindle (2) consisting of microtubules. Note that the mitochondria remain intact during mitosis. Magnification 14,000×.

Histology

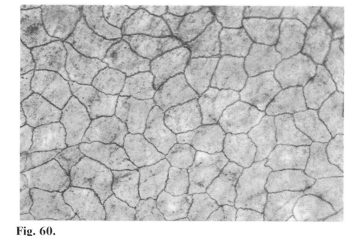

Fig. 60.

Goblet cell

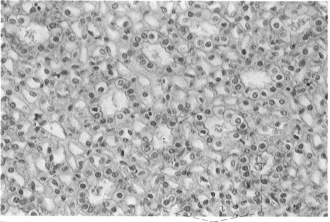

Fig. 61.

Lumen of collecting tubule

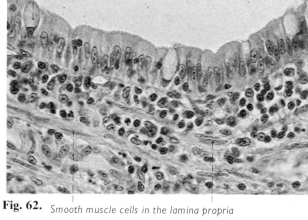

Fig. 62. *Smooth muscle cells in the lamina propria*

Fig. 60. *En face* view of a thin spread of a cat peritoneum treated with silver nitrate. The cells of this simple squamous epithelium are clearly demarcated by silver deposits at the cellular interfaces. No counterstaining for the nuclei. Magnification 240×.

Fig. 61. Simple cuboidal epithelium lining the collecting tubules in the renal medulla (rabbit). The thyroid follicles are also used to demonstrate this type of epithelium (cf. Fig. 409). Mallory-azan staining. Magnification 240×.

Fig. 62. Simple columnar epithelium from the cat's jejunum with goblet cells (i.e., unicellular glands). The striated border is visible at the luminal surface of the absorptive epithelial cells, but is seen better at higher magnification (cf. Fig. 73). Note smooth muscle cells in the lamina propria. H.E. staining. Magnification 380×.

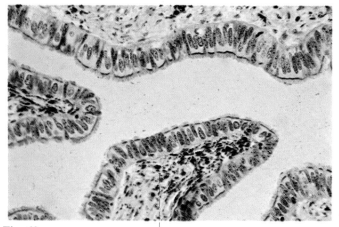

Fig. 63. *Connective tissue in fold of mucous membrane*

Fig. 63. Simple ciliated columnar epithelium lining the mucosal folds of the human uterine tube. The narrow black line at the base of the cilia would be resolved at higher magnification to consist of many fine dots corresponding to the basal bodies. Iron-hematoxylin staining. Magnification 240×.

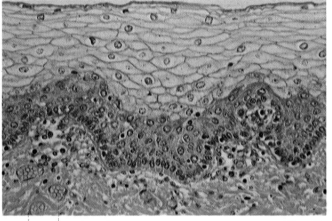

Venules stuffed with erythrocytes   **Fig. 64.**

Fig. 64. Human vaginal epithelium that belongs to the non-keratinizing variety of stratified squamous epithelia. All cells including those in the superficial layer contain nuclei, but only the latter show the characteristic squamous shape. Therefore the classification of stratified epithelia is based on the shape of the cells found in the uppermost layer! Goldner staining. Magnification 240 ×.

Cornified layer

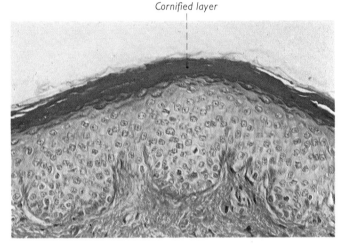

Fig. 65.

Fig. 65. Slightly keratinized stratified squamous epithelium from the skin of human nostrils (cf. Fig. 421). The surface cells are non-nucleated and become completely transformed into horny plates. Mallory-azan staining. Magnification 240 ×.

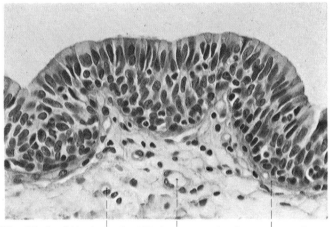

Fig. 66. Small blood vessels within lamina propria   Basement membrane

Fig. 66. Stratified columnar epithelium (very rare) of human female urethra with only the cells in the uppermost layer giving a columnar appearance. But because it is the shape of these cells that determines the formal classification of stratified epithelia, it is "columnar." Mallory-azan staining. Magnification 380 ×.

Epithelia – Pseudostratified

Goblet cells

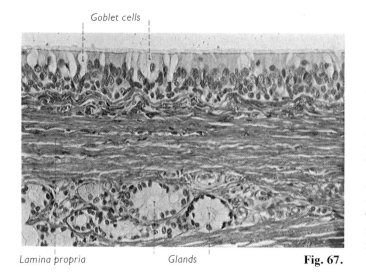

Lamina propria   Glands   **Fig. 67.**

Fig. 67. Ciliated pseudostratified columnar epithelium from human trachea with goblet cells (note cell shape in the superficial layer!). As this variety of epithelium occurs only in the respiratory tract, it is often referred to as "respiratory epithelium." It is classified as "pseudostratified" because all the cells rest on the basement membrane, but not all of them reach the free surface. These details, however, are obscured in most of the common histological specimens as in this figure. The student will learn, with growing experience, that what might look like a "stratified" ciliated columnar epithelium on first inspection in reality always is a "pseudostratified" one. Mallory-azan staining. Magnification 240×.

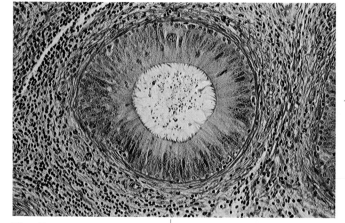

Fig. 68.   Basement membrane

Fig. 68. Pseudostratified columnar epithelium with stereocilia from human ductus epididymidis. In contrast to the motile cilia, the basal bodies are missing and the stereocilia of each cell are stuck together at their free ends (see also Fig. 75). When viewed under the electron microscope, the stereocilia are seen to consist of long branching cell processes lacking the characteristics of cilia. Iron-hematoxylin and benzo light bordeaux staining. Magnification 150×.

Surface cells with superficial layer of condensed cytoplasm

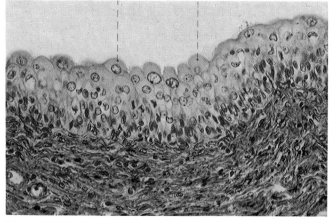

Fig. 69.

Fig. 69. Transitional epithelium from human urinary bladder. The appearance of this epithelium varies greatly depending upon the state of contraction or distention of the hollow organs in which it is found. In routine histologic specimens it usually appears as "stratified"; and the student should, therefore, take into account "transitional epithelium" when dealing with stratified epithelia in general. Characteristic features of the transitional epithelium are the surface cells that often are binucleated with a superficial layer of condensed darker staining cytoplasm (see also Fig. 72). Mallory-azan staining. Magnification 240×.

Terminal bars

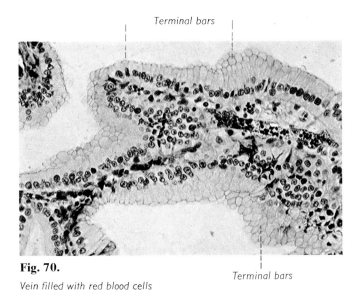

Fig. 70.

Fig. 70. Mucosal fold from human gall bladder covered with a high simple columnar epithelium. When sectioned tangentially a hexagonal array of blackish lines surround the apical portions of the cells. These are referred to as "terminal bars" that consist at the level of the electron microscope of a sequence of three different attachment devices forming the junctional complex (cf. Fig. 49b). Iron-hematoxylin staining. Magnification 240×.

Terminal bars

Vein filled with red blood cells

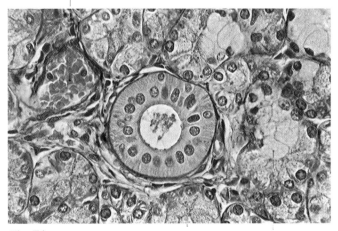

Fig. 71.

Fig. 71. Striated (salivary) duct from human submandibular gland. The basal portions of the columnar cells present a striated appearance because their mitochondria are oriented perpendicularly to the base of the cell. The electronmicroscopic equivalent of this basal striation is called "basal labyrinth" and is shown in Fig. 19. This is a particularly well-differentiated Mallory-azan staining, as can be seen from the orange-yellow shade of the erythrocytes (cf. Fig. 2). Mallory-azan staining. Magnification 380×.

Serous and mucous alveolus

Binucleated surface cell   Crusta

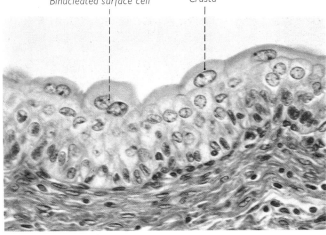

Fig. 72.

Fig. 72. Transitional epithelium from human urinary bladder. A unique specialization of this epithelium is the occurrence of a zone of condensed and darker staining cytoplasm subjacent to the adluminal surface. It contains a mixture of different glycoproteins and displays many filaments and membrane-bound vesicles at the level of the electron microscope. Mallory-azan staining. Magnification 380×.

Basement membrane

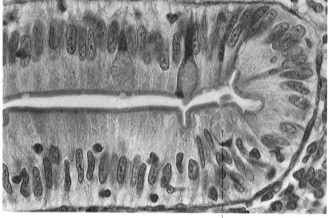

Fig. 73.

Goblet cell with nucleus

> The three most common specializations found at epithelial surfaces are represented by cell processes of different shape and structure.

Fig. 73. This micrograph of human intestinal epithelium illustrates the striated (brush) border (stained a pale grayish violet) which is particularly well developed in all kinds of absorptive cells. Electron micrographs reveal that it is composed of numerous regularly arranged microvilli of uniform height (cf. also Fig. 18). Mallory-azan staining. Magnification 600 ×.

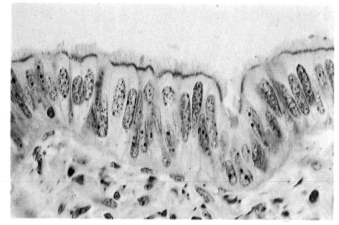

Fig. 74.

Fig. 74. In contrast the simple columnar epithelium of the human fallopian tube possesses motile cilia at its surface. These can easily be distinguished from either striated border or stereocilia by means of their basal bodies, that in this case show as a narrow bluish-black line. Due to the poor staining properties cilia are often overlooked, but become clearly visible by refraction when the iris diaphragm of the microscope condensor is closed. Iron-hematoxylin staining. Magnification 600 ×.

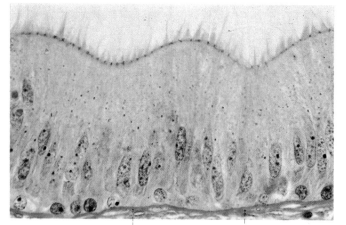

Fig. 75.

Basement membrane

Fig. 75. Photomicrograph of human ductus epididymidis. The nonmotile stereocilia of each cell surface are stuck together at their free ends and lack basal bodies. Electron micrographs show that they are unusually long branching cell processes. The extremely fine dark dots located between the apical ends of the epithelial cells correspond to cross sections of the terminal bars (see also Fig. 70). Hematoxylin and benzo light bordeaux staining. Magnification 600 ×.

44

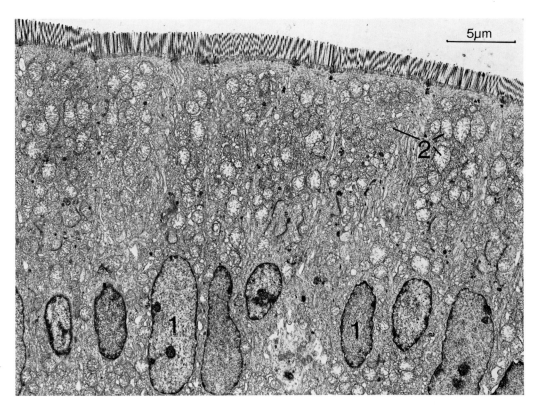

Fig. 76. Simple columnar epithelium from rat intestine with closely and regularly spaced microvilli of uniform size and shape (cf. Fig. 73). The supranuclear parts of the cells are crowded with mitochondria (2) and membrane profiles preferably belonging to the smooth ER. Due to the low resolution of this micrograph no other structural details can be seen. 1 = Nucleus. Magnification 3,500×.

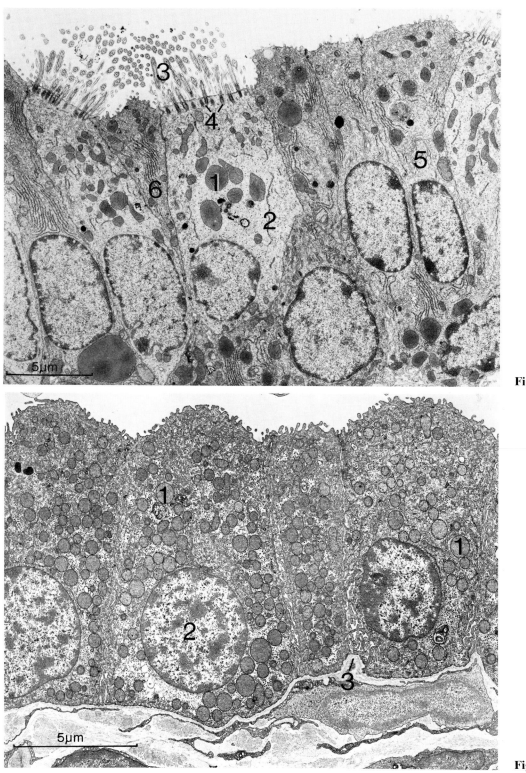

Fig. 77.

Fig. 78.

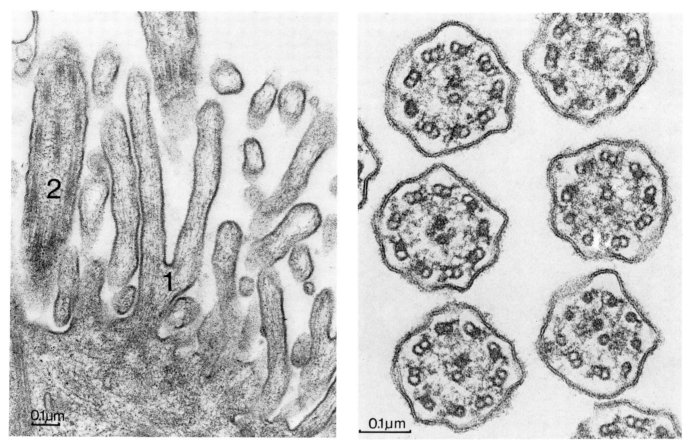

Fig. 79a.

Fig. 79b.

Fig. 79. a) Arborizing and irregularly outlined microvilli (1) projecting together with kinocilia (2) from the same epithelial cell of a feline bronchus. Magnification 70,000×.

b) Transverse sections through kinocilia from feline respiratory epithelium. Note the trilaminar appearance of the outer ciliary membrane and the complex internal structure. This consists of two single microtubules in the center of the axoneme encircled by nine uniformly spaced groups of paired microtubules (doublets); and hence this arrangement, universal in all kinocilia, is often referred to as the "9 + 2" structure. Magnification 130,000×.

◀

Fig. 77. Pseudostratified columnar epithelium with several ciliated cells of the mouse trachea. The ciliated cells appear lighter in the electron microscope, and they contain a few profiles of both smooth and rough ER (2) together with mitochondria that vary considerably in size (1). The kinocilia (3) are cut transversely, obliquely and longitudinally, yet reveal nothing of their delicate internal structure. Merely their basal bodies (kinetosomes) are clearly outlined due to their high electron density (4). The other and darker cell type, so-called Clara cells, shows prominent Golgi complexes (5) together with regularly arrayed stacks of the rough ER (6). The apical pole of these cells is crowded with tubules representing the smooth ER that becomes apparent with higher resolutions (cf. Fig. 25). Magnification 4,500×.

Fig. 78. Simple low columnar epithelium from feline respiratory bronchiole. The cell surfaces show short and irregular microvilli; the cell body is homogeneously filled with secretion granules (1) yet it reveals no more details of its fine structure as, e.g., a well-developed smooth ER due to the rather low magnification. 2 = Nucleus; 3 = Basal lamina. Magnification 6,500×.

47

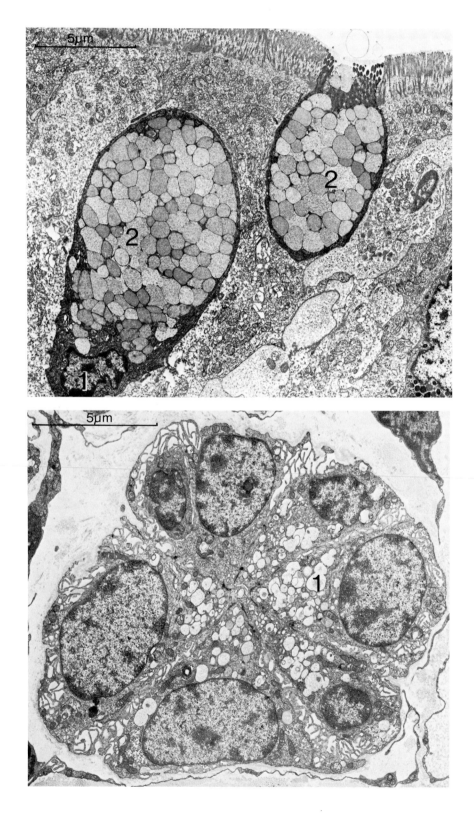

Fig. 80. Two goblet cells located within the intestinal epithelium of a rat. (Compare with the light micrographs of Figs. 73 and 83.) While the left of these two glandular cells displays parts of its nucleated bottom that also comprises most of the cytoplasm, the other shows its opening fringed by microvilli. 1 = Nucleus; 2 = Secretion (mucus) granules. Magnification 4,000×.

Fig. 81. Transverse section of a serous acinus found in the submucosa of a small feline bronchus. Note the characteristic cuneate outline of the secretory cells, the roundish nuclei located at the cell base and the extremely narrow slit-shaped lumen. 1 = Secretion granules. Magnification 6,500×.

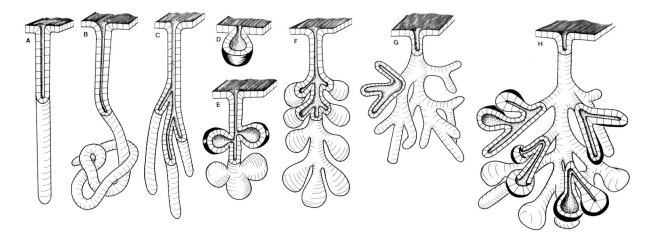

Fig. 82. Schematic diagram for the classification of exocrine glands based on the different shapes of their secretory units and the arrangement of their duct system. A) Simple tubular gland, i.e., each secretory unit opens separately on the epithelial surface (e.g., crypts of the colon). B) Simple coiled tubular gland (e.g., sweat glands of the skin). C) Simple branched tubular gland, i.e., several secretory units join in a single unbranched secretory duct (e.g., glands in the pyloric mucosa of the stomach). D) Simple alveolar gland. E and F) Simple branched alveolar glands (e.g., sebaceous glands of the skin). G) Compound tubular gland, i.e., the tubular secretory units open into an elaborate and branched duct system. All these varieties also occur in alveolar and acinar glands and finally a combination within the same gland of tubular and alveolar/acinar secretory units is possible (H). In case of the latter the different secretory portions either follow one after the other, thus forming "mixed tubuloacinar or tubuloalveolar" glands (e.g., the submandibular and sublingual glands), or the different secretory units remain separate and are not connected with each other and a "tubuloacinar" or "tubuloalveolar" gland results. In all these last-mentioned cases we are confronted with compound glands, i.e., glands with an intensely branching duct system. For classification of exocrine glands cf. Table 5.

Nucleus of a goblet cell

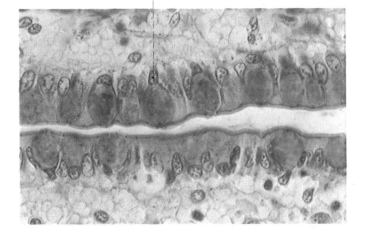

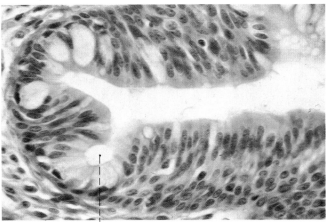

Lumen of intraepithelial gland

Fig. 83. Goblet cells in the epithelial lining of the ileum as an example of a unicellular, intraepithelial gland. With the Mallory-azan staining all mucous secretory granules are stained brilliant blue. Note the cuneiform nucleus located in the "stem" of the goblet and the prominent striated border of the epithelium. Mallory-azan staining. Magnification 600×.

Fig. 84. Multicellular intraepithelial gland in the mucosa of the human nasal septum. Iron-hematoxylin and benzo purpurin staining. Magnification 380×.

Glandular epithelia – Different forms of secretory units

Lumen of epithelial crypt

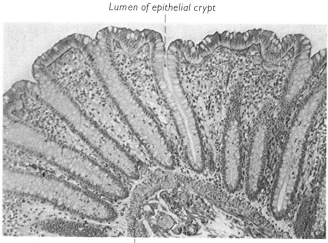

Fig. 85.

Muscularis mucosae

Fig. 85. The crypts of the colic mucosa (man) represent the classic example of simple tubular glands as their walls are composed mainly of secretory (goblet) cells. As the infoldings of the crypts are not oriented exactly perpendicular to the free surface, they are often cut tangentially with only fragments lying in the plane of the section. Note cross-sectioned smooth muscle of the muscularis mucosae at the base of the crypts. Mallory-azan staining. Magnification 95×.

Lumen of two serous acini

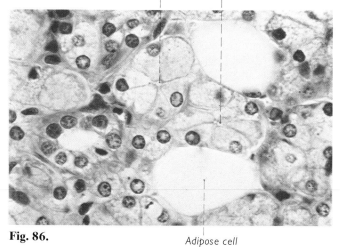

Fig. 86.

Adipose cell

Fig. 86. Transverse section of a secretory unit is clearly seen in the center of this micrograph (human parotid gland). Note the cuneiform shape and the globular nuclei of the secretory cells that border upon a narrow lumen that is particularly prominent in this case (cf. Fig. 81). Mallory-azan staining. Magnification 600×.

Myoepithelial cells

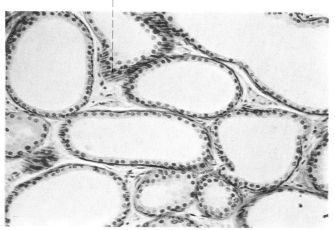

Fig. 87.

Fig. 87. Extremely wide lumen of alveolar secretory units of the ceruminal glands from human external auditory canal. Because of their secretion mechanism they are designated with the misleading name: large apocrine "sweat glands." Mallory-azan staining. Magnification 150×.

Pyloric pit continuing into a simple branched tubular gland

Duct of sebaceous gland

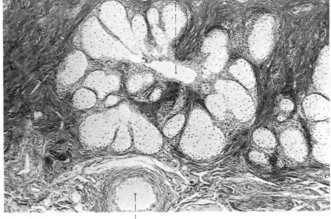

Fig. 88.

Fig. 89.

Small artery

Fig. 88. Simple branched tubular glands from human pyloric mucosa. One has to search the specimen thoroughly to see their tubular shape and branching sites, because these lie only here and there within the plane of the section. H.E. staining. Magnification 60×.

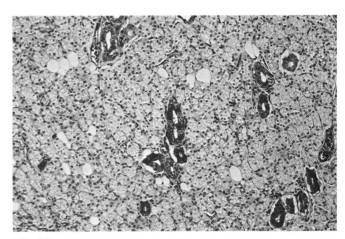

Fig. 90.

Fig. 89. Simple branched alveolar gland (sebaceous gland from human upper eyelid). The lumen of the secretory portions are not visible because 1. most of them are sectioned tangentially and 2. the holocrine secretion mechanism gradually transforms the secretory cells into the secretory product that fills the lumen. Mallory-azan staining. Magnification 60×.

Fig. 90. Serous gland (human parotid gland) that, according to the form of its secretory portions, is to be classified as an "acinar" gland and whose elaborate and branched ducts clearly demonstrate that it is a "compound" gland. Mallory-azan staining. Magnification 96×.

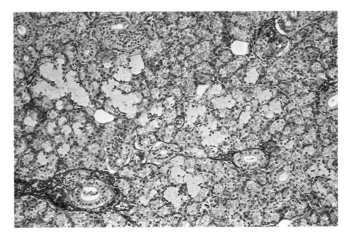

Fig. 91. Human submandibular gland that can be classified 1. according to the nature of its secretion as a "mixed" gland; 2. according to the form of its secretory units, as a "tubulo-acinar" gland and 3. due to its elaborate and branching duct, as a "compound" gland. The mucous cells form the tubular parts of the secretory portions and are stained a brilliant blue. Mallory-azan staining. Magnification 96×.

Fig. 91.

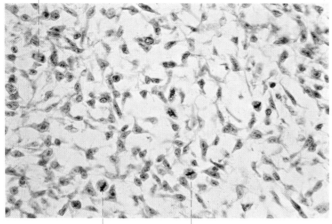

Late anaphase

Fig. 92. *Metaphases*

Fig. 92. Mesenchyme from the head region of a chick embryo. The hollow spaces between the stellate cells are filled with a succulent, amorphous ground substance. No fibers are visible. Iron-hematoxylin staining. Magnification 380×.

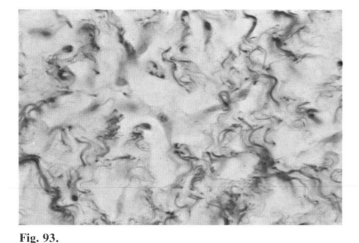

Fig. 93.

Fig. 93. Mucous connective tissue with fibers (Wharton's jelly of human umbilical cord). Along with a few fibroblasts a network of fine collagenous fibers is seen in the amorphous ground substance. Mallory-azan staining. Magnification 380×.

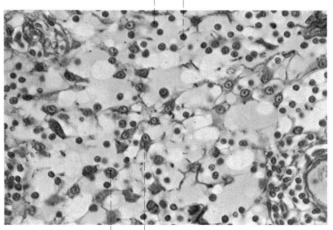

Lymphocytes

Fig. 94. *Reticular cells*

Fig. 94. Reticular connective tissue from medullary sinus of a cat lymph node. In the center of the micrograph a network of stellate reticular cells can be seen with delicate reticular fibers (stained brilliant blue) closely attached to their surfaces and with their many processes resembling mesenchymal cell. The round, apparently "naked" nuclei belong to lymphocytes. Mallory-azan staining. Magnification 380×.

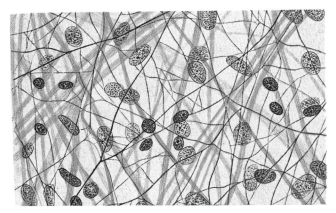

Fig. 95.

Collagenous fiber

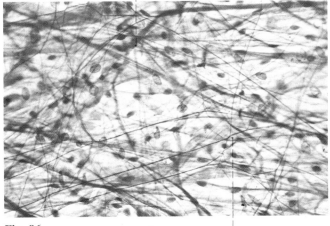

Fig. 96.
Branching site of an elastic fiber
Branching elastic fiber

Fig. 95 and **96.** Moderately schematized drawing (Fig. 95) and a light micrograph (Fig. 96) of the same specimen, a spread of connective tissue from rat omentum showing collagenous and elastic fibers. The collagenous fibers run straight in this preparation instead of their usual undulating course. Contrary to the elastic fibers they never branch but are interwoven, forming some sort of a loosely arranged wickerwork, while the delicate elastic fibers do branch thus forming a true network (cf. Table 7). The "naked" nuclei belong to the various types of connective tissue cells. Hornowsky's staining (combination of resorcin-fuchsin and van Gieson). As this stain is not very stable and tends to fade, both fiber types are similar in color. Magnification 240×.

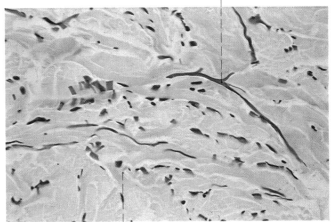

Fig. 97.
Bundle of collagenous fibers

Fig. 97. Collagenous and elastic fibers of human subcutaneous connective tissue. The broad collagenous fibers staining a pale brown are intersected in all directions by a very fine network of elastic fibers (cf. Table 7). The various types of connective tissue cells cannot be seen because nuclear counterstaining was not performed. Resorcin-fuchsin staining. Magnification 240×.

Fig. 98. Reticular fibers demonstrated in the liver by the silver impregnation technique (hence their name "argyrophilic" fibers). Because of their optical and physicochemical properties these fibers range between their collagenous and elastic counterparts. They are arranged as a delicate filigree-like meshwork along the interface between the interstitial connective tissue (= stroma) and the specific cells (= parenchyma) of each organ, thus forming a mold of the contents they enclose, i.e., the cords of hepatic cells. Bielschowsky's staining. Magnification 240×.

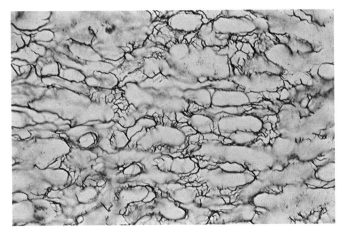

Fig. 98.

Connective tissue – Cell types

Mast cell

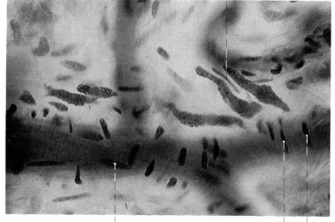

Fig. 99.   Endothelial nucleus   Nuclei of media cells

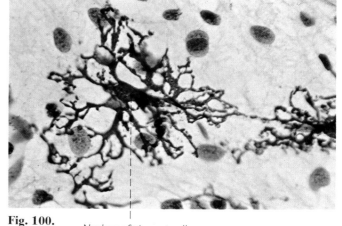

Fig. 100.   Nucleus of pigment cell

Serous alveolus

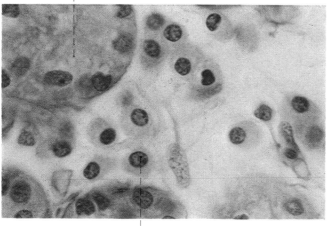

Fig. 101.   Plasma cell

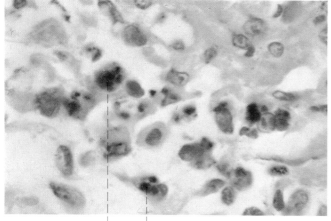

Fig. 102.   Histiocytes

Fig. 99. A spread of canine isolated periosteum showing mast cells along a small artery. The mast cell granules are stained metachromatically due to their high content of heparin, a mucopolysaccharide. Blood vessels injected with a blue gelatine solution. Toluidine-blue staining. Magnification 600×.

Fig. 100. Richly arborized pigment cell in the subcutaneous tissue of a salamander larva containing large numbers of dark brown or black melanin granules, hence called "melanocyte." Hematoxylin staining. Magnification 380×.

Fig. 101. Plasma cells from the interstitial connective tissue of a human lacrimal gland. A characteristic feature of these cells is their round, eccentrically placed nucleus. The "cartwheel" appearance resulting from regularly arranged chromatin particles within the nucleus is seen much more rarely than is generally believed. The basophilia of the protoplasm (hence the gray-blue tinge with the azan stain) is due to the abundance of RNA in the form of bound ribosomes on the rough-surfaced endoplasmic reticulum (cf. Fig. 112). Mallory-azan staining. Magnification 960×.

Fig. 102. Histiocytes "labeled" by the incorporation of the vital stain trypan blue (loose connective tissue from mouse skin). The high phagocytic activity allows for a selective demonstration of these motile cells that because of this ability, belong to the reticuloendothelial system. Vital staining with trypan blue, nuclear fast red. Magnification 960×.

Small artery

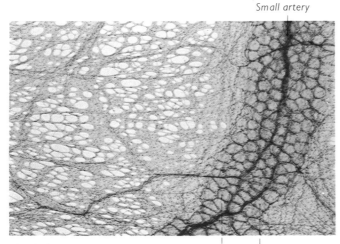

Fig. 103.

Adipose tissue

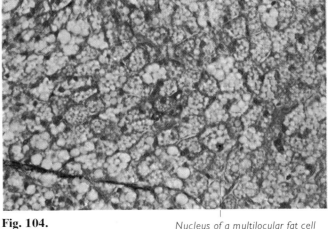

Fig. 104.

Nucleus of a multilocular fat cell

Adipose cell

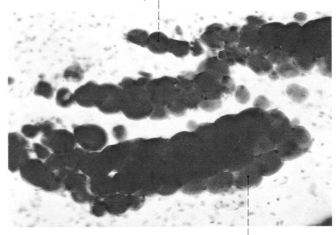

Fig. 105.

Adipose cell

Fig. 103. Real areolar connective tissue is only represented in the major and minor omentum and some of the mesenteries. The many elliptical and empty spaces do not correspond to fat cells, but represent true cavities surrounded by a connective tissue framework that is covered by a mesothelial lining. On the right side of the micrograph a branching artery embedded in adipose tissue can be seen. Hematoxylin and benzo light bordeaux staining. Magnification 38×.

Fig. 104. Multilocular adipose tissue from a cat fetus. As the name indicates, each cell contains several fat droplets that gradually coalesce, finally resulting in one big droplet that nearly completely fills the cell body. Also in this case many of the nuclei are already pushed to the periphery, but they still preserve their globular shape. Mallory-azan staining. Magnification 240×.

Fig. 105. Small lobules of adipose tissue from rat mesentery. The individual fat cells are often ill defined because they are too closely attached to and upon each other. In this preparation the lipid droplet that almost entirely fills the cell body has been preserved and is selectively stained. Hemalum and lipid red staining. Magnification 150×.

Fig. 106. Adipose tissue from a human lacrimal gland. As the fat is dissolved out in all paraffin – or celloidin – embedded material while the tissue is dehydrated by alcohol, benzene, etc., the cells appear as "empty" profiles. The boundaries of these do not correspond to the cell membrane, but consist of the plasmalemma together with a thin fringe of cytoplasm displaced to the cell periphery by the large fat droplets. The nuclei are flattened and usually found pressed against the cell wall. Mallory-azan staining. Magnification 150×.

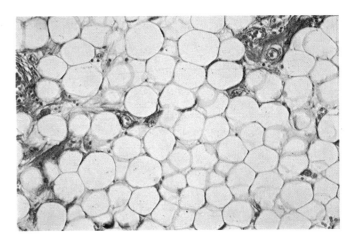

Fig. 106.

55

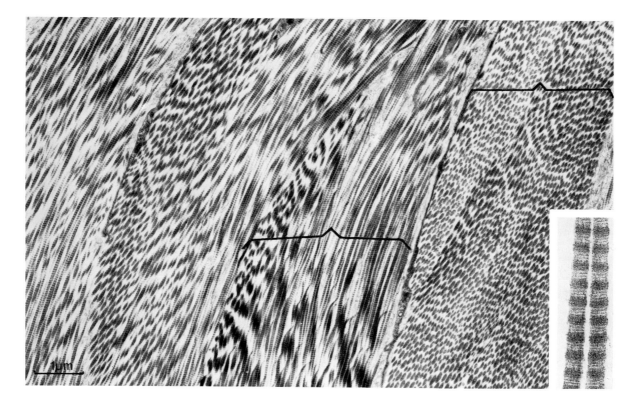

Fig. 107.

Fig. 107. Collagenous fibers from rat subcutaneous connective tissue. As these fibers usually crisscross each other under variant angles they are found in routine electron micrographs in all kinds of sections from exactly transverse to various degrees of oblique to exactly longitudinal. Irrespective of the rather low resolution, a coarser periodicity consisting of darker and lighter bands can be seen in the longitudinally sectioned fibrils. This cross striation can be further resolved into smaller zones of alternating electron density (inset). The brackets mark the width of one collagenous fiber. Magnification 12,000 × and 56,000 ×.

Fig. 108. Richly vascularized loose connective tissue from feline submandibular gland showing numerous slender and elongated cytoplasmic processes (1) resembling a whiplash that only occasionally exhibit their continuity with the cell body of a fibroblast (2_1, 2_2). 3_1 and 3_2 = Lumen of postcapillary venules; 4= Bundle of cross-sectioned collagenous fibrils. Magnification 8,000 ×.

Fig. 109a and **b.** Parts of the perikarya of a fibrocyte (a) and a fibroblast (b) that contain the nuclei and the majority of the organelles. The fibroblast can be identified as such by its well-developed and evidently highly active rough endoplasmic reticulum (1) together with the numerous Golgi complexes (2) indicative of synthetic processes. 3 = Unmyelinated nerve. Magnification 14,500 × and 10,500 ×.

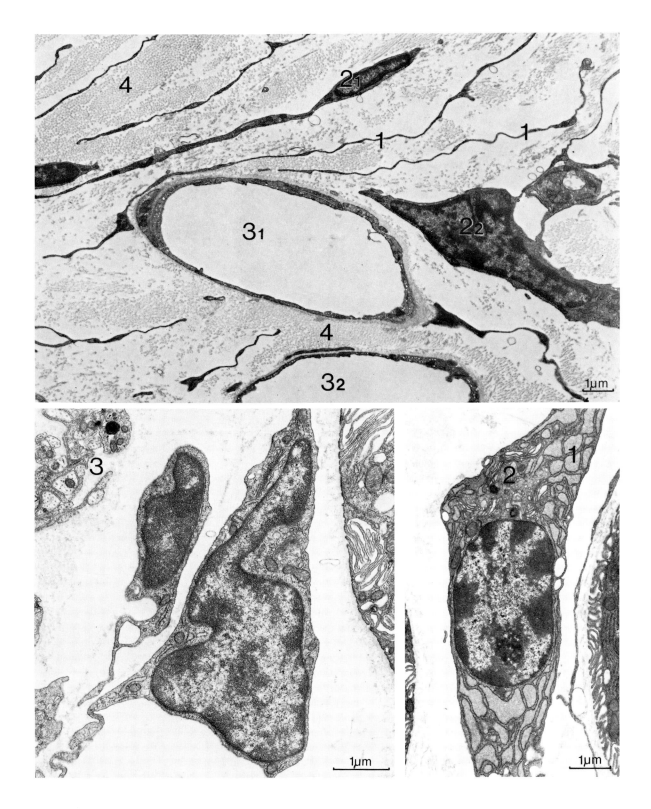

Fig. 108.

Fig. 109.

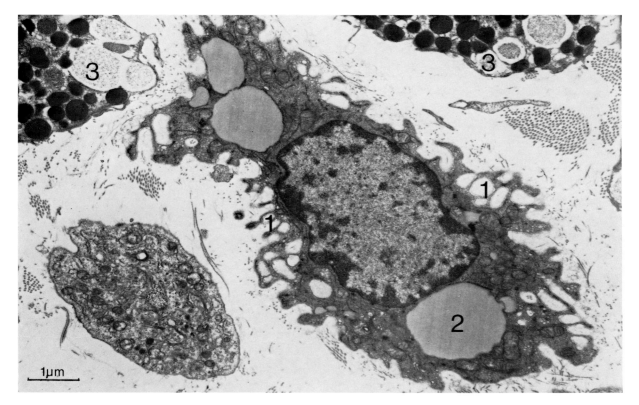

Fig. 110. Histiocyte from rat subcutaneous connective tissue. The fine structure of these ameboid and highly phagocytic cells is characterized by an intensely vacuolated outer cytoplasmic fringe (1) that corresponds to the many pseudopod-like processes. Besides the usual organelles, the cell body may contain a large variety of inclusion bodies originating from the enzymatic breakdown of phagocytized materials into secondary lysosomes and residual bodies. However, in this case the histiocyte only contains lipid droplets (2) of different sizes. 3 = Parts of a mast cell. Magnification 13,000×.

Fig. 111. Mast cell from pulmonary loose connective tissue (cat). Characteristic for this type of mobile cells are the large roundish and membrane-bound granules of variant electron density (1, 2, 3). These corespond to various stages of maturation of these corpuscles that contain histamine and heparin. Magnification 18,000×.

Fig. 112. Plasma cell from the submucosa of the rat duodenum that belong to the facultative cellular components of the loose connective tissue. Plasma cells are characterized by a well-developed ergastoplasm serving for the synthesis of immune globulins. This becomes also apparent here in the form of the dilated cisternae (*) of the ergastoplasm that contain a material (protein) of moderate electron opacity. Magnification 19,500×.

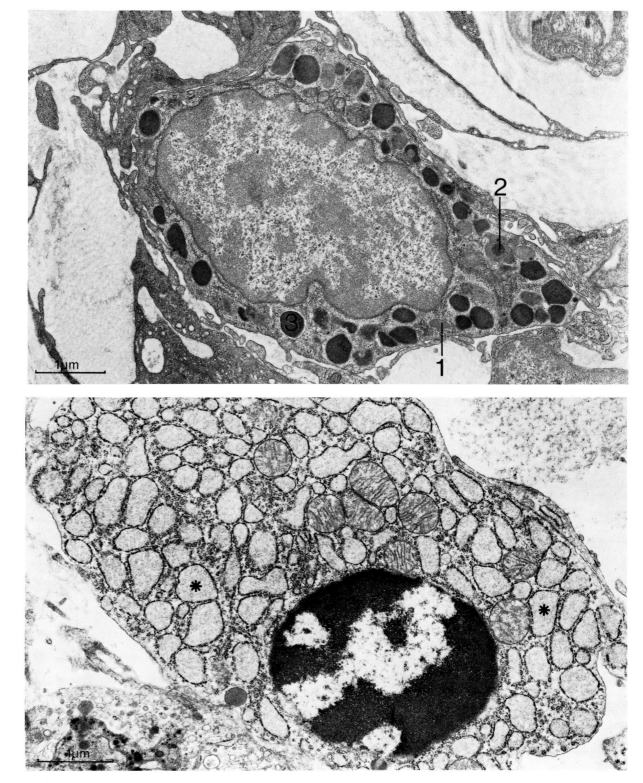

Fig. 111.

Fig. 112.

Loose connective tissue

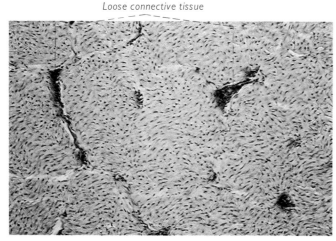

Fig. 113.

Strand of loose connective tissue

Fig. 114.

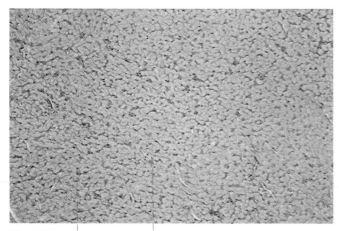

Fig. 115. *Nuclei of fibroblasts*

Fig. 116.

Fig. 113. Cross-sectioned canine tendon clearly exhibiting its subdivision into smaller fiber bundles by strands of loose connective tissue. Note large number of fibroblast nuclei that can be found in nearly every interstice between the tendon fibers. This observation is confirmed and is, perhaps, even more evident when this section is compared with a longitudinal section (cf. Fig. 114). H.E. staining. Magnification 95×.

Fig. 114. Longitudinal section of the same tendon as shown in Fig. 113. Note fibroblasts arranged in alternating parallel rows with only their nuclei being visible. The upper part of the micrograph is traversed by a strand of loose connective tissue. The undulating appearance is one characteristic of tendon fibers which, however, can also be found in longitudinally sectioned nerve trunks and hence is no criterion for the final identification of the tissue. H.E. staining. Magnification 95×.

Fig. 115. Cross section of an elastic ligament, the bovine nuchal ligament. In this photomicrograph the elastic fibers are stained green (with azan or with eosin-methylene blue they would stain a brilliant red) and the sparse and delicate collagenous fibers are stained blue to bluish green. The collagenous fibers are homogeneously distributed. The nuclei present throughout the section are predominantly those of fibroblasts which are found in association with both the elastic and collagenous elements (cf. Fig. 113). Iron-hematoxylin and picro-indigocarmine staining. Magnification 95×.

Fig. 116. Longitudinal section of the same specimen as shown in Fig. 115. Note the small numbers of nuclei and the broad and partially parallel elastic fibers that branch frequently and fuse at acute angles as in a stretched fishing net (cf. Fig. 114). Iron-hematoxylin and picroindigocarmine staining. Magnification 95×.

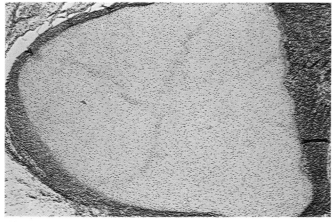

Fig. 117.

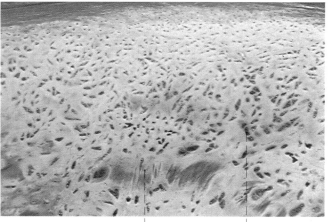

Fig. 118.   *Amianthoid degeneration*   *Groups of chondrocytes*

Fig. 117. Fetal hyaline cartilage from human calcaneus. Note that the many cartilage cells lie separately throughout the tissue instead of being arranged in groups and that the matrix appears homogeneous and stains uniformly (cf. Fig. 118). Mallory-azan staining. Magnification 38×.

Fig. 118. Mature hyaline cartilage from human rib at a low magnification showing the cells being arranged in groups and the matrix staining differently, demonstrating an heterogeneous distribution of different components of the intercellular materials. In the lower part of the micrograph a typical degenerative alteration of hyaline cartilage can be seen. By changing the chemical composition of the ground substance the collagenous fibers become visible, replacing the matrix by their tightly packed bundles. This is known as amianthoid degeneration. H.E. staining. Magnification 38×.

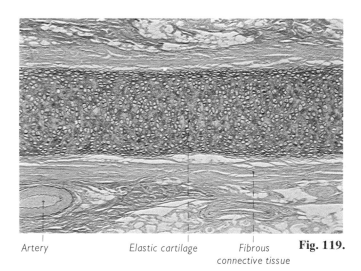

Artery   *Elastic cartilage*   *Fibrous connective tissue*   **Fig. 119.**

Fig. 119. Elastic cartilage from pig's external ear. The cartilage cells are scattered regularly throughout the matrix and are often found in groups of two which are not often seen in hyaline cartilage. The intercellular material is stained a dark violet due to a selective staining of the elastic network by resorcin-fuchsin. The individual elastic fiber cannot be visualized because of the low magnification (cf. Fig. 123). Resorcin-fuchsin staining, nuclear fast red. Magnification 38×.

Fig. 120. Fibrous cartilage from human intervertebral disk. In this type of cartilage the collagenous fibers are unmasked and hence they are always visible, being quite often arranged in a characteristic herring-bone pattern. The matrix contains only a few, mostly singly distributed cells that can hardly be recognized at this low magnification. H.E. staining. Magnification 38×.

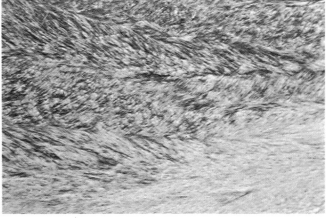

Fig. 120.

Connective tissue – Cartilage

Epithelium of bronchus

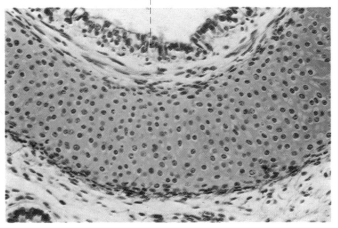

Fig. 121.

Territorial matrix   *Amianthoid degeneration*

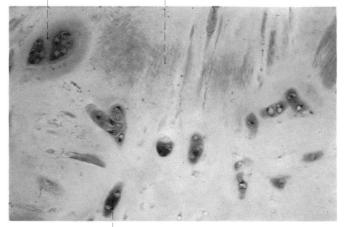

Fig. 122.   *Two chondrocytes in territorial matrix*

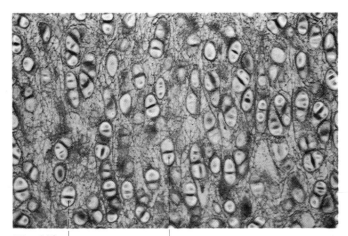

Fig. 123. *Bicellular groups of chondrocytes*

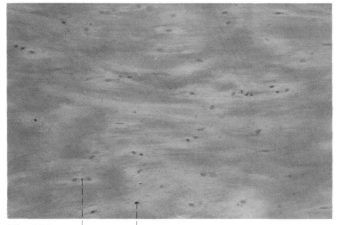

Fig. 124. *Cartilage cells in hyaline matrix*

Fig. 121. Hyaline cartilage from the bronchus of a human fetal lung. The singly distributed chondrocytes show a nearly circular outline (note shape of nucleus) and are embedded in a homogeneous matrix. Mallory-azan staining. Magnification 240×.

Fig. 122. Mature hyaline cartilage from human rib showing amianthoid degeneration (in the upper parts of the micrograph) and groups composed of rather small cartilage cells (same specimen as in Fig. 118). The translucent spaces correspond to the lacunae in which the cells are located. Due to histological processing the cells shrink considerably into small masses with often only their nuclei being visible as intensely staining but artificially altered corpuscles. The dark basophilic areas surrounding the cell groups are known as "capsules" that are parts of the territorial matrix and particularly rich in chondromucoprotein. H.E. staining. Magnification 150×.

Fig. 123. Elastic cartilage from pig's external ear (same specimen as in Fig. 119) whose chondrocytes are considerably less shrunken than those shown in Fig. 122. Therefore, their nuclei (stained a faint pink) preserved their spherical shape surrounded by the weakly staining halo of the cell body. Due to shrinkage these are separated from the walls of the lacunae. Note the bicellular groups of chondrocytes and the dense network of delicate elastic fibers. Resorcin-fuchsin staining, nuclear fast red. Magnification 150×.

Fig. 124. Fibrocartilage from human intervertebral disk. The uni- or bicellular groups of chondrocytes are irregularly distributed between the collagenous fiber bundles of the matrix, with only their nuclei clearly recognizable. H.E. staining. Magnification 150×.

Hair germs   Epidermis   Blood vessel

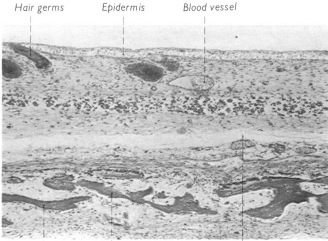

Intramembranous bone   Cleft caused by shrinkage   **Fig. 125.**

Osteoblasts

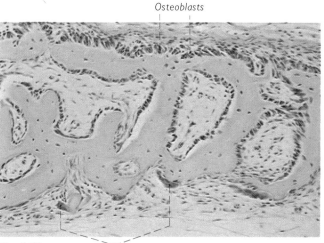

Fig. 126.

Osteocytes
Osteoblasts

Fig. 127.   Osteoclasts

Osteoblasts

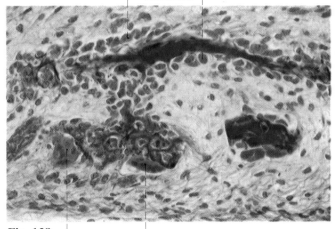

Fig. 128.   Osteoclasts in Howship's lacunae

Fig. 125. Human fetal cranium, as an example of intramembranous bone development. In this case mesenchymal cells transform into osteoblasts that produce the noncalcified osseous ground substance, the osteoid. Some of the osteoblasts become surrounded by the osteoid matrix and, with the deposition of calcium salts into the osteoid, become osteocytes as the osteoid hardens progressively around them. H.E. staining. Magnification 38 ×.

Fig. 126. A close-up of Fig. 125 (from its lower right corner) clearly illustrates the osteoblasts closely attached to the surface of the osseous trabecula and the osteocytes located in their interior. H.E. staining. Magnification 150 ×.

Fig. 127. Trabeculae from canine fetal mandible. The trabecular surfaces – particularly those oriented toward the skin – are covered with osteoblasts (serving as the producers of osteoid), whereas the osseous surfaces facing the oral cavity show osteoclasts (multinuclear giant cells) that are responsible for the enzymatic resorption of bone. H.E. staining. Magnification 95 ×.

Fig. 128. Osseous trabeculae (stained blue) from porcine fetal skull covered by numerous osteoblasts that are responsible for the growth of bone by means of apposing more and more ground substance. At the same time enzymatic resorption of bone is occurring on the inner surface by the activities of the osteoclasts. Small recesses (Howship's lacunae) are formed in which these cells are often located. The apposition of bone on the outer surface accompanied by the resorption of bone on the inner surface results in the enlargement of the cranium and accommodates the rapidly developing brain. Mallory-azan staining. Magnification 240 ×.

Fig. 129. Early stage of intracartilaginous osteogenesis in the maniphalanx of a three-month-old human fetus. In contrast to the intramembranous bone development in this form of ossification, there first appears a cartilaginous model of the later bone that is reabsorbed and then gradually replaced by osseous tissue. The entire process starts with a calcification of the cartilaginous ground substance near the center of the future shaft, the diaphyseal or primary ossification center. This is accompanied by proliferation and hypertrophy of the chondrocytes and by the deposition of a bony collar around the cartilage of the ossification center. As the latter is derived from osteoblasts that develop from the mesenchymal cells of the perichondrium, this location of bone formation ist known as perichondral ossification. But note that, notwithstanding these different names and terms, the basic mechanisms by which osseous tissue is formed are identical in both intramembranous and intracartilaginous ossification (camera lucida drawing). H.E. staining. Magnification 80 ×.

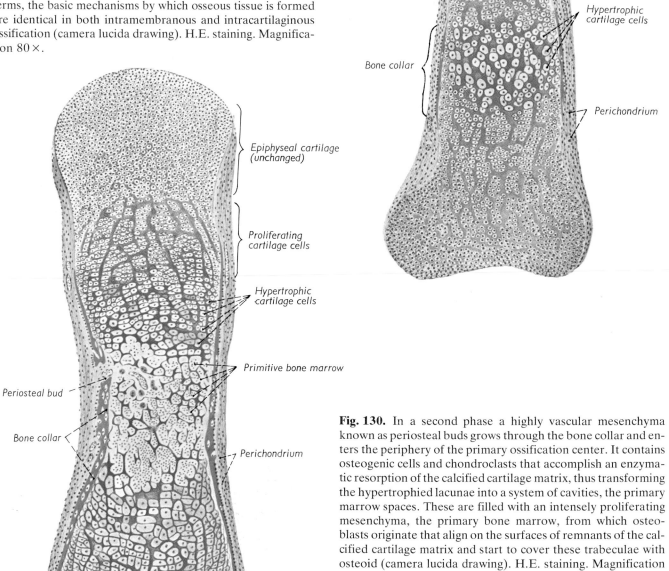

Epiphyseal cartilage
(unchanged)

Proliferating
cartilage cells

Hypertrophic
cartilage cells

Primitive bone marrow

Periosteal bud

Bone collar

Perichondrium

Hypertrophic
cartilage cells

Bone collar

Perichondrium

Fig. 130. In a second phase a highly vascular mesenchyma known as periosteal buds grows through the bone collar and enters the periphery of the primary ossification center. It contains osteogenic cells and chondroclasts that accomplish an enzymatic resorption of the calcified cartilage matrix, thus transforming the hypertrophied lacunae into a system of cavities, the primary marrow spaces. These are filled with an intensely proliferating mesenchyma, the primary bone marrow, from which osteoblasts originate that align on the surfaces of remnants of the calcified cartilage matrix and start to cover these trabeculae with osteoid (camera lucida drawing). H.E. staining. Magnification 100 ×.

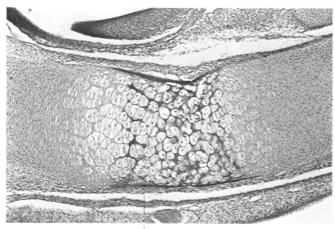

Fig. 131.   *Perichondral bone*

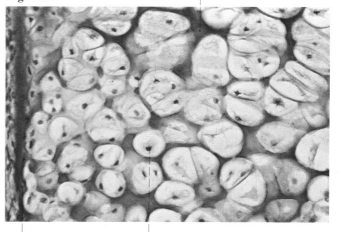

Fig. 132.   *Calcified cartilagineous ground substance*

Fig. 131. Longitudinal section through a metacarpal bone from a human fetus with hypertrophied chondrocytes and calcification of the intervening cartilage matrix that thereby becomes more basophilic. This process establishes the primary ossification center. H.E. staining. Magnification 60×.

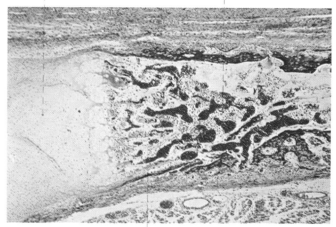

Perichondral bone   *Nucleus of a shrunken chondrocyte*
Epiphyseal cartilage   *Primitive bone marrow*

Fig. 132. A close-up of the primary ossification center from the preceding micrograph clearly shows the enlargement of the lacunae that is caused by a considerable swelling of the chondrocytes. This sign of cellular degeneration, however, has been lost due to the shrinkage that often occurs in routine histologic preparations. H.E. staining. Magnification 240×.

Fig. 133.   *Perichondral bone*

Fig. 133. Longitudinal section through a metatarsal bone from an 18 cm human fetus that matches the camera lucida drawing of Fig. 130. Mallory-azan staining. Magnification 38×.

Blood vessels stuffed with erythrocytes

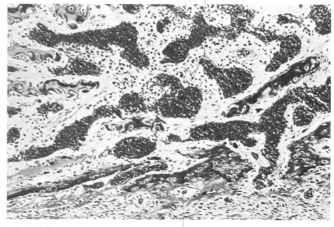

Fig. 134. A close-up of the lower parts of the middle third of the preceding micrograph reveals numerous openings in the periosteal bone collar through which mesenchyma together with blood vessels enters the primary marrow cavity. Mallory-azan staining. Magnification 96×.

Fig. 134.   *Perichondral bone*

Connective tissue – Development and growth of bone

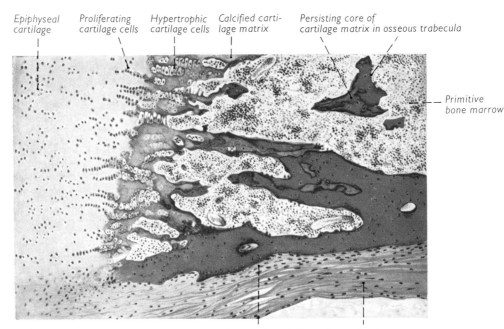

Epiphyseal cartilage | Proliferating cartilage cells | Hypertrophic cartilage cells | Calcified cartilage matrix | Persisting core of cartilage matrix in osseous trabecula

Primitive bone marrow

Osteogenic and fibrous layer of periosteum

Fig. 135. Detail from a third and later stage of intracartilaginous bone development. The marrow cavity has enlarged considerably toward the epiphyseal ends, where it reaches the hyaline cartilage that shows calcification of the matrix and therefore a stronger basophilia together with a hypertrophy of the cells. Note that here the same degenerative processes precede the resorption of cartilage as those already described for the formation of the primary ossification center (see Fig. 129). Remnants of the calcified original cartilage ground substance allow for an early attachment of osteoblasts and thus serve as "guidelines" for the ossification and as a supporting framework that persists for some time (camera lucida drawing). H.E. staining. Magnification 80 ×.

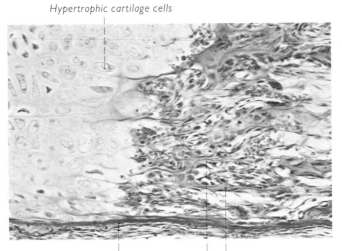

Hypertrophic cartilage cells

Perichondral bone | Spicules of calcified cartilage matrix

Fig. 136. Border zone between epiphyseal cartilage and marrow cavity from the proximal third of a human fetal humerus. Remnants of the calcified cartilage matrix (colored violet) project like spicules from the zone of hypertrophied cartilage into the highly cellular and intensely vascularized marrow cavity. H.E. staining. Magnification 150 ×.

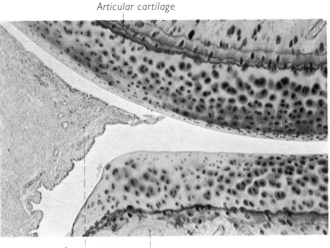

Articular cartilage

Synovial fold | Bone

Fig. 137. Detail of a section from a human fetal knee joint whose articular cartilage is a persisting derivative of the former epiphyseal cartilage that in this location is already clearly divided into territories. At the left side of the micrograph note synovial villus protruding into the joint cavity. H.E. staining. Magnification 38 ×.

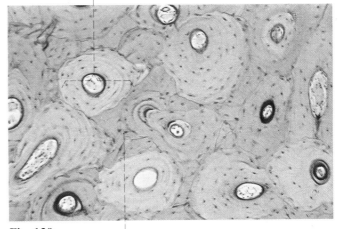

Haversian system (= osteon)

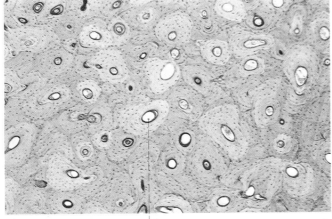

Fig. 138. Haversian canal

Fig. 139. Interstitial lamellae

Fig. 138 and **139.** Transverse section through the compact bone of a human fibula showing numerous Haversian systems, each of which consists of several osseous lamellae concentrically arranged around a circular opening, the Haversian canal. A higher resolution (Fig. 139) reveals the osteocytes as darker colored dots wedged in between the osseous lamellae and aligned into concentrical circular lines. Remnants of former Haversian systems constitute the interstitial lamellae filling the spaces between the osteons. Fuchsin staining. Magnification 38× and 96×.

Fig. 140. Haversian canal

Fig. 140. Longitudinal section through the compact bone of a canine humerus. In the center of the micrograph a Haversian canal is cut along its total length. In contrast to the ground-bone preparation as shown in Fig. 141, in true sections through decalcified bone remnants of the periosteum and the loose connective tissue in the Haversian canals can always be seen. Carbol-thionine staining. Magnification 38×.

Fig. 141. Ground-bone preparation from the shaft of a canine femur. When these paper-thin- slices – prepared by grinding down a piece of bone by means of abrasives – are transferred into a staining solution, the almond-shaped cavities and the delicate canaliculi, in which the osteocytes together with their processes were originally located, are clearly outlined. The lacunae are always arranged parallel to the osseous lamellae situated between or within the latter. The canaliculi penetrate the lamellae at right angles as shown. Fuchsin staining. Magnification 240×.

Lacuna Canaliculi

Fig. 141. Haversian canal

Muscular tissue

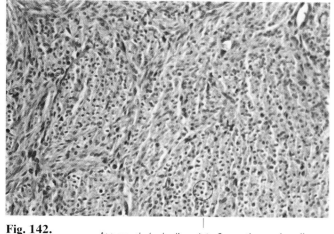

Fig. 142.

Apparently 'naked' nuclei of smooth muscle cells
Myofibrillae-free sarcoplasmic area

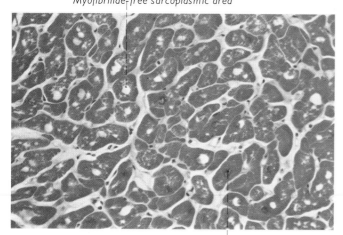

Fig. 143.

Nucleus of cardiac cell
Nucleus of skeletal muscle fiber

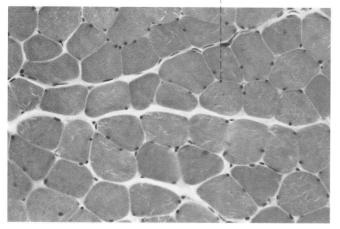

Fig. 144.

'Red' muscle fiber

Fig. 145.

'White' muscle fiber

Fig. 145. Demonstration of three different types of skeletal muscle fibers by means of their variant contents of glycogen (tibialis ant. muscle, rat). The "red" fibers with an extremely low concentration of glycogen appear nearly unstained, the intermediate type fibers are moderately colored and the "white" fibers rich in glycogen take a brilliant magenta stain. PAS staining. Magnification 96 × (specimen courtesy of Dr. U. Osterkamp).

Fig. 142–144. The three types of muscular tissue illustrated in transverse sections with the same stain (H.E.) and with identical magnifications (× 240). Irrespective of the fact that the nuclei are not very prominent at this rather low magnification, the different sizes of the cross-sectional areas of muscle *cells* (human uterine smooth muscle, Fig. 142 and canine myocardium, Fig. 143) and of muscle *fibers* (human sternohyoideus muscle, Fig. 144) are a helpful criterion in distinguishing the various muscle tissues one from another. Compare with the matching longitudinal sections in Figs. 147–149. The "empty" spaces in many of the cardiac cells correspond to the axial sarcoplasmic areas free of myofibrils that are located at the nuclear poles.

Fig. 146.
Erythrocytes within a capillary

Fig. 146. Special stains like iron-hematoxylin are often used to demarcate the cross striation particularly clearly, thereby allowing for a definite identification of the A- and I-bands even at rather low magnifications (canine hyoidal muscle). The bluish-black corpuscles aligned parallel to the edges of the fibers are not the muscle fiber nuclei but represent erythrocytes that are remarkably compressed within the narrow blood capillaries. Iron-hematoxylin staining. Magnification 240×.

Small bundle of smooth muscle cells

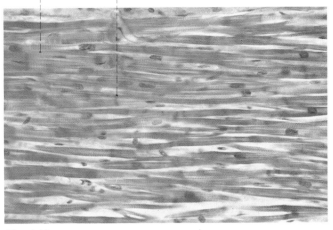

Fig. 147.
Small vein

Cross striation   *Intercalated disk*

Fig. 148.

Fig. 147–149. The three types of muscular tissue illustrated in longitudinal sections with the same stain (H.E.) and with identical magnifications (×240). Note the axial position of the nucleus in both smooth muscle (human myometrium, Fig. 147) and cardiac cells (canine myocardium, Fig. 148), while in muscle fibers the nuclei are located directly subjacent to the sarcolemma; and, in addition, they occur in very large numbers within one fiber (canine hyoidal muscle, Fig. 149). Both the cross striations and the intercalated discs are barely visible in this preparation.

Nuclei of skeletal muscle fibers

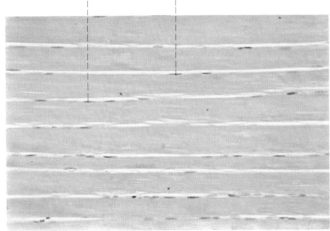

Fig. 149.

Muscular tissue

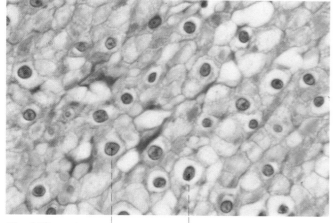

Fig. 150.

Nuclei of smooth muscle cells

The three different structural elements found in muscle tissues demonstrated in transverse sections with the same stain (Mallory-azan) and with identical high resolution (× 100 oil immersion objective) and magnification (× 960) for a better identification of cytological details.

Fig. 150. Smooth muscle cells from the muscular tunic of a human vermiform appendix. Note the axial position of the nuclei surrounded by a small rim of cytoplasm.

A single cardiac cell

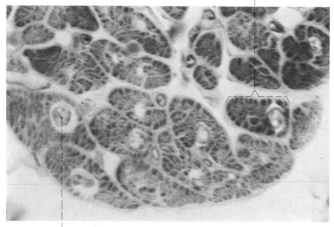

Fig. 151. *Nucleus surrounded by myofibrillae-free sarcoplasm*

Fig. 151. Myocardial cells display a remarkably larger yet more irregularly outlined cross-sectional area relative to smooth muscle cells. Their myofibrils are often arranged in small groups, known as Cohnheim fields, separated by narrow strands of sarcoplasm.

Erythrocyte within a capillary

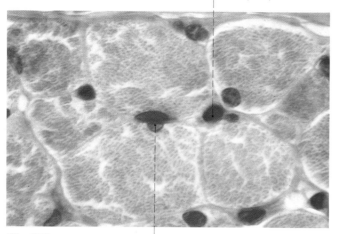

Fig. 152.

Nucleus of skeletal muscle fiber

Fig. 152. Clearly visible myofibrils in skeletal muscle fibers from human orbicularis oculi muscle.

The three different structural elements found in muscle tissues demonstrated in longitudinal sections with the same stain (Mallory-azan) and with identical high resolution (× 100 oil immersion objective) and magnification (× 960) for a better identification of cytological details.

Nucleus of smooth muscle cell

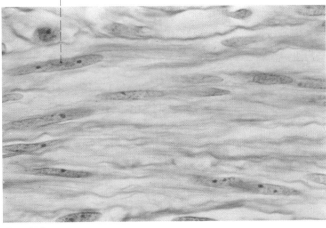

Fig. 153.

Fig. 153. Smooth muscle cells from the muscular tunic of a human vermiform appendix. While the rod-like nuclei with their prominent nucleoli are clearly visible, the cell boundaries remain poorly defined. This is usually found in longitudinal sections of smooth muscle due to the delicacy of its cells.

Intercalated disks

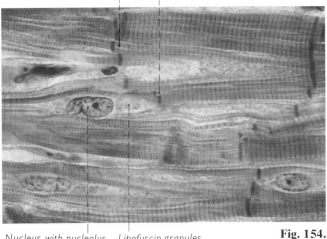

Nucleus with nucleolus *Lipofuscin granules* **Fig. 154.**

Fig. 154. Myocardial cells from a dog clearly exhibit sarcoplasmic areas at the nuclear poles that are free of myofibrils, but contain lipofuscin granules. Note both the intercalated disks (darker red) that serve for the cohesion of successive cells and the cross and longitudinal striations of the cells that depend on the myofibrils.

A-band *Z-line*

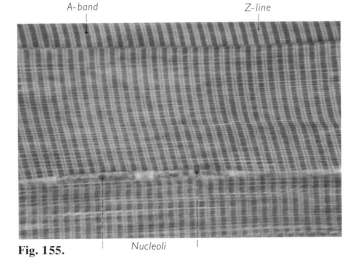

Fig. 155. Skeletal muscle fibers with prominent cross striation from human pectoralis major muscle. As the lighter I-(= isotropic) and the darker A-(= anisotropic)bands are almost of identical width, the fibers must have been fixed in a largely relaxed state. Note the prominent Z-lines in the middle of each of the I-bands, while the M- and H-bands cannot be identified within the A-bands. Directly subjacent to the sarcolemma two nuclei are faintly visible, in particular their nucleoli.

Fig. 155. *Nucleoli*

Muscular tissue

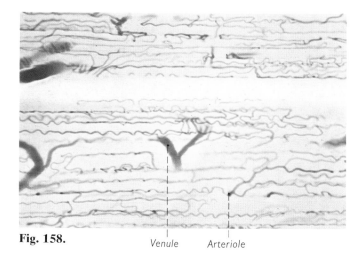

Fig. 156.

Nuclei of arborizing smooth muscle cells

Adipose cell

Fig. 158.   Venule   Arteriole

Fig. 156. Arborizing smooth muscle cells with prominent triangular nuclei in a thin spread of a frog's urinary bladder. The rest of the muscle cells are compressed into bundles consisting of extremely long and slender elements that barely disclose their nuclei. H.E. staining. Magnification 240×.

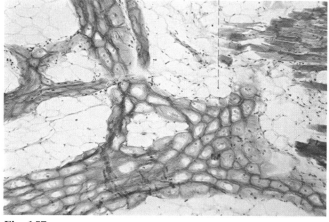

Fig. 157a.

Fig. 158. Vascular injection specimen of a canine skeletal muscle clearly demonstrating the patterns of its microvasculature. The wavy, tortuous course of the capillaries is neither due to the contraction of the muscle fibers nor is it a means to accommodate to variations in fiber length, but it is a characteristic feature of capillaries supplying "red" fibers. Injection with a gelatine solution colored with carmin, no counterstain. Magnification 96×.

Ordinary cardiac muscle

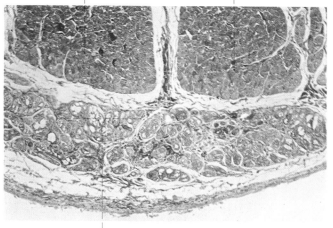

Fig. 157a, b. Purkinje fibers of the heart (dog and man) shown in longitudinal and transverse section with the same stain (Mallory-azan) and with identical magnification (×96). These ultimate branches of the impulse conducting system are composed of highly specialized cardiac cells that can be distinguished from the ordinary myocardial cells by their larger size, their high content of glycogen and their reduced amount of myofibrils that are shifted to the cell periphery. In addition, the small round nuclei are distinctive for the Purkinje fibers, but they are met less frequently by the sections than those of the ordinary cardiac cells due to their small size relative to the large volume of the cells proper.

Small bundle of Purkinje fibers   **Fig. 157b.**

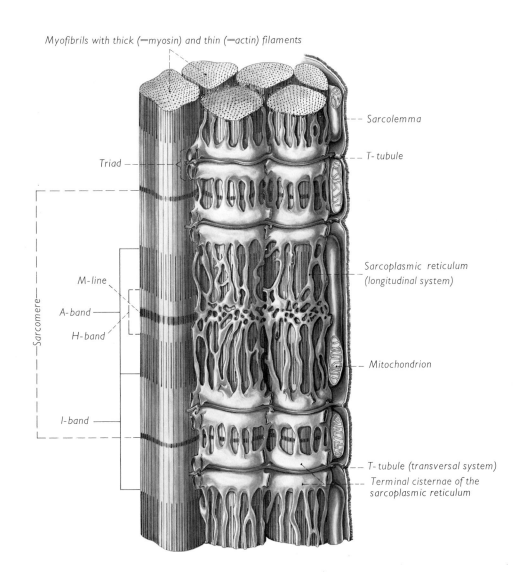

Myofibrils with thick (=myosin) and thin (=actin) filaments

Sarcolemma

T-tubule

Triad

Sarcomere

M-line

A-band

H-band

I-band

Sarcoplasmic reticulum (longitudinal system)

Mitochondrion

T-tubule (transversal system)

Terminal cisternae of the sarcoplasmic reticulum

Fig. 159. Schematic tridimensional representation of the interrelationships of the sarcoplasmic reticulum (= longitudinal system), the transverse tubules (T-tubules) and the myofibrils in a mammalian skeletal muscle fiber. The triads of the reticulum consist of a centrally positioned T-tubule flanked by two terminal cisternae of the sarcoplasmic reticulum that appear as translucent vacuoles in ultra-thin sections. In mammalian muscles the triads are located at the A-I junctions, hence there are two triads to each sarcomere [redrawn and extensively modified after Peachey: J. Cell Biol. 25, 29 (1965)].

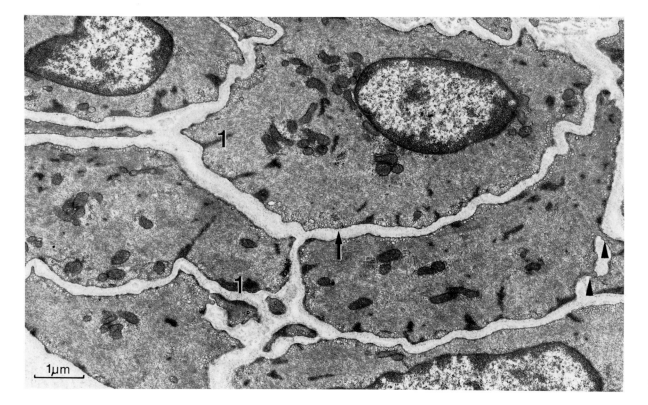

Fig. 160.

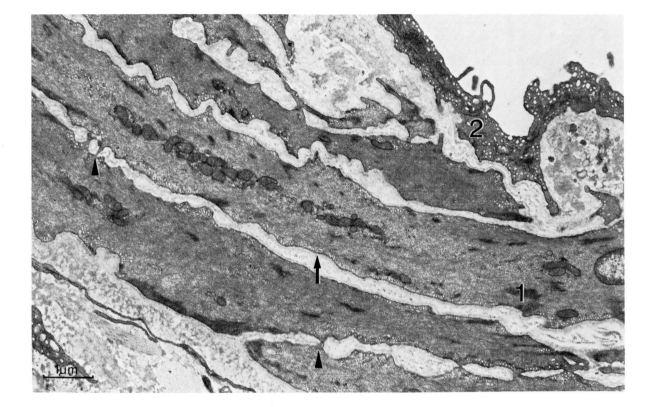

Fig. 161.

Fig. 160, 161. Electron micrographs of smooth muscle cells from the media of a feline small artery. The size and appearance of these cells vary within wide ranges depending on the level where they are cut and on the number and shape of their outward projections. The cell body is almost homogeneously filled with a filamentous material of uniform density (myofilaments) so that the mitochondria together with the rest of the organelles occur preferably close to the nuclei. The oval or fusiform densities, so-called dense spots (1), with a prevailing localization along the inner aspect of the cell membrane, serve as attachment sites for the contractile substance. In addition, the muscle cells make numerous close contacts by means of small pleomorphic projections (▶) that display gap junctions (= nexus) at their cellular interface. These are known to be sites of low electric resistance permitting the intramural spread of excitation from one cell to another. The membrane vesicles or caveolae either arranged in rows or small groups (⟶) serve as a store for calcium ions, and they are therefore and because of their close interrelationship with the poorly developed smooth ER compared with the T-tubules to which they should correspond. 2 = Endothelium. Magnification 13,000×.

Cardiac muscle – Electron microscopy

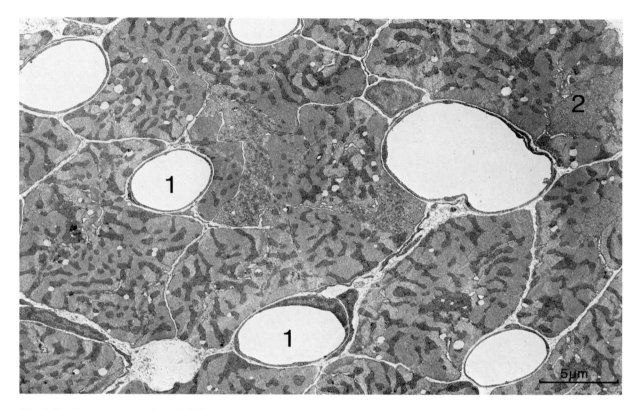

Fig. 162. Transverse section of feline papillary muscle displaying the characteristic differences in size and shape of the individual cross-sectioned cardiac cells that are often but poorly defined. Note the close interrelationship between capillaries (1) and muscle cells. 2 = Nucleus of a cardiac cell. Manification 4,000×.

Fig. 163. Oblique section through the myocardium of a pig clearly showing the A- and I-bands together with the Z-lines of its cross-banded pattern. Also here the individual cells are difficult to distinguish one from another due to their close apposition, but the capillaries (1) are helpful in detecting the narrow interstitial spaces. 2 = Nucleus of a cardiac cell. Magnification 4,000×.

Fig. 164. Parts of two cardiac cells (papillary muscle, pig) of which the lower one shows several myofibrils cut longitudinally, thereby disclosing that their cross-banded pattern is almost indentical to that of skeletal muscle (see p. 79). 1 = Endothelial nucleus; 2 = Capillary lumen. Magnification 25,000×.

Fig. 163.

Fig. 164.

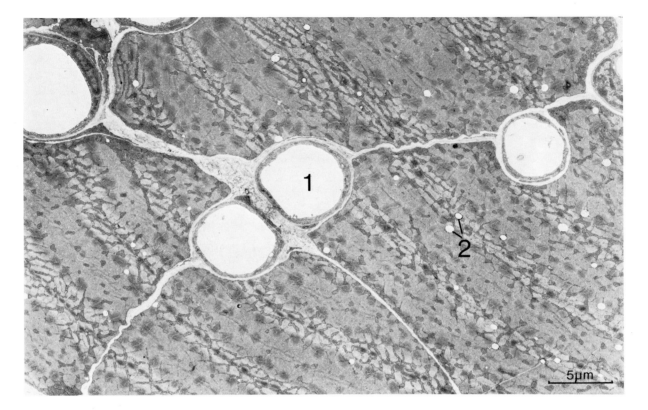

Fig. 165. Cross-sectioned skeletal muscle fibers from a feline tongue. When comparing this micrograph with Fig. 162 it becomes apparent that the same area comprises merely parts of four muscle fibers, while it contains at least seven entire cardiac cells. This gives a rough estimate of the order of magnitude by which skeletal muscle fibers usually differ from both smooth and cardiac muscle cells. 1 = Capillarly lumen; 2 = Lipid vacuoles. Magnification 3,500×.

▶

Fig. 166. Longitudinal section through skeletal muscle fibers from the same specimen and shown with the same magnification as in the preceding electron micrograph. As the prominent Z-lines are flanked by very narrow I-bands these fibers are remarkably contracted. Note the strands of mitochondria (→) located between the myofibrils. 1 = Capillary lumen; 2 = Lipid vacuoles; 3 = Arteriolar lumen; 4 = Nucleus of a muscle fiber. Magnification 3,500×.

Fig. 167. Transverse section of a myofibril at the level of the A-I junction encircled by the T-tubule (→). Compare with Fig. 159. Note the hexagonal array of the thick myosin filaments between which the thin actin filaments are but poorly defined. 1 = Part of a nucleus; 2 = Part of a mitochondrium. Magnification 78,000×.

Fig. 168. Two moderately contracted myofibrils cut longitudinally, clearly displaying all components of their cross-banded pattern (cremaster muscle, rat). Particularly note the darker M-band (1) bisecting the lighter H-band located in midposition of the A-band (cf. Fig. 159). The H-band is that central part of the A-band into which the actin filaments do not reach, while the M-band is due to delicate filamentous cross bridges between the individual myosin filaments. At the A-I junctions triads are visible in cross (2) and oblique sections (3) that consist of three closely attached cavities. The most central and minute lumen belongs to the T-tubule that is flanked by two larger vacuoles that correspond to cross-sectioned terminal cisternae of the sarcoplasmic reticulum (see also the tridimensional representation of Fig. 159). 4 = Z-line. Magnification 26,000×.

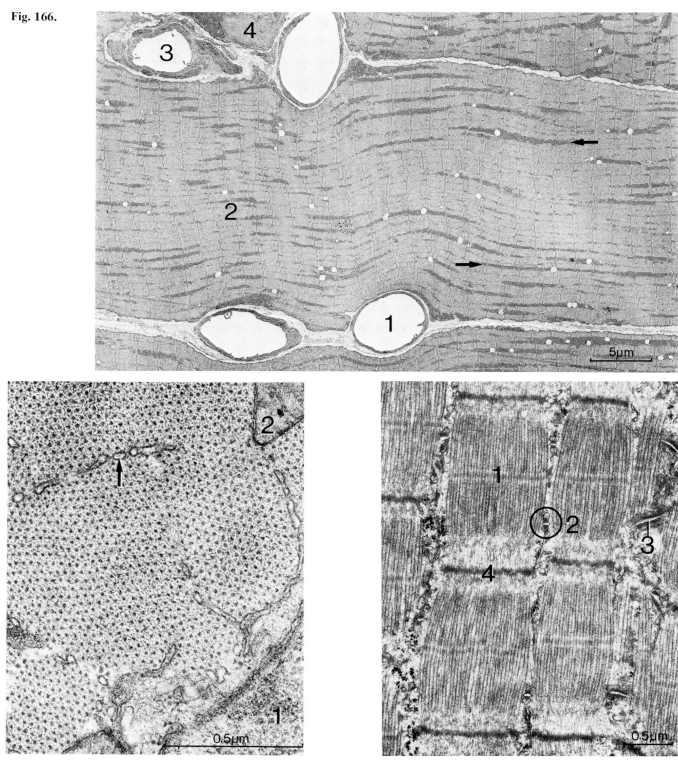

Fig. 166.

Fig. 167.

Fig. 168.

Nervous tissue – Nerve cells

Glia cell nuclei

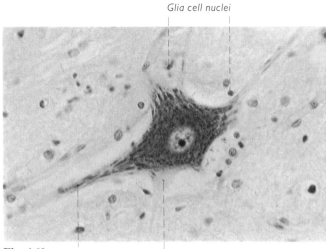

Fig. 169. *Dendrite   Axon hillock*

Pyramidal cells

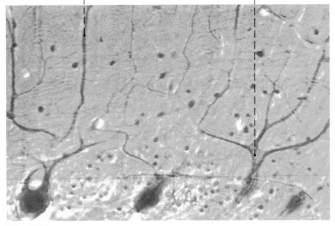

Fig. 170. *Multipolar nerve cell*

Dendrites of Purkinje cells

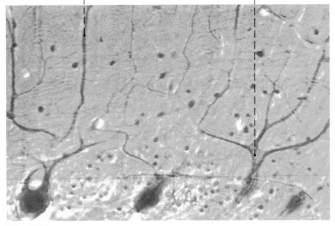

Fig. 171.

Fig. 169. Multipolar nerve cell from the anterior horn of the canine spinal cord showing several dendrites that can be identified by the regular occurrence of Nissl bodies in their proximal parts, while the axon and the region of the cell body from where it emerges, the axon hillock, are conspicuously free of Nissl substance. Note the large spherical nucleus with its prominent nucleolus. The strongly basophilic clumps of material are termed Nissl substance according to their first disoverer, and they represent the light microscopical equivalent of a well-developed rough-surfaced endoplasmic reticulum. Toluidine-blue staining. Magnification 380×.

Fig. 170. Several pyramidal and one multipolar nerve cell in human cerebral cortex. Due to the complex tridimensional distribution of their cytoplasmic processes, often only the bodies (= perikarya) of the arborized ganglion cells can be seen in routine histologic sections. Note the remarkable diversity in size exhibitied by nerve cells that are shown with the same magnification in Figs. 169, 170 and 172. Silver impregnation (Schultze-Stöhr). Magnification 380×.

Fig. 171. The fan-shaped dendritic arborization of a Purkinje cell from rat cerebellar cortex. The axon is given off from the end of the flask-like cell bodies opposite the dendrites. Silver impregantion (Schultze-Stöhr). Magnification 240×.

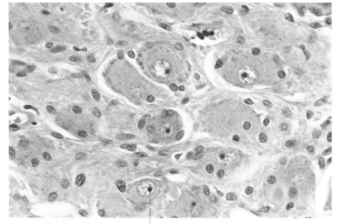

Fig. 172. *Nucleus with nucleolus*

Fig. 172. Multipolar nerve cells in an autonomous ganglion found in the human adrenal medulla. The majority of these cells appear as globular elements, and their large vesicular nuclei together with their prominent nucleoli allow for a distinct identification of these cells. Mallory-azan staining. Magnification 380×.

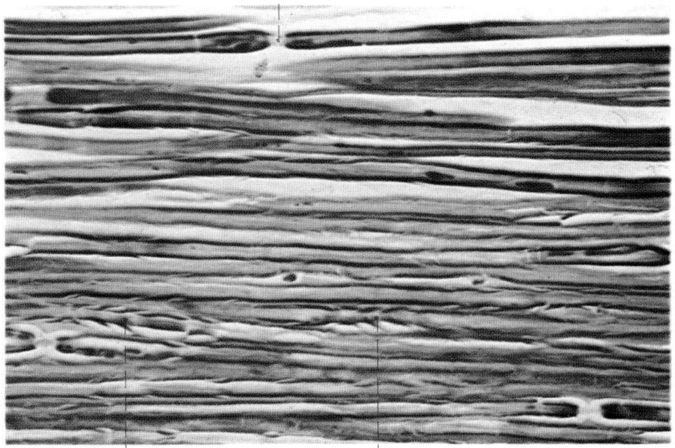

Node of Ranvier

Fig. 173. Longitudinal section of rabbit ischiadic nerve fixed in OsO₄ which preserves and blackens myelin. A node of Ranvier (= interruption of the myelin sheath) can be seen in the upper and lower part of the micrograph together with funnelshaped incisures, the clefts of Schmidt-Lanterman. The latter represent focal areas where the myelin lamellae are separated by Schwann cell cytoplasm bur retain their continuity. Fixation in OsO₄, no counterstain. Magnification 240×.

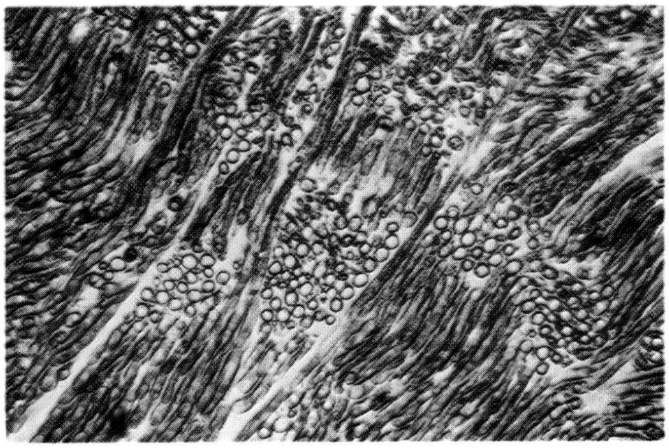

Fig. 173. Schmidt-Lanterman clefts

Fig. 174. Feline spinal nerve in cross and longitudinal section. When cross-sectioned, the myelin sheaths appear as darkbrown rings encircling an unstained an hence "faint" central core, the axon. Fixation in OsO₄; no counterstain. Magnification 150×.

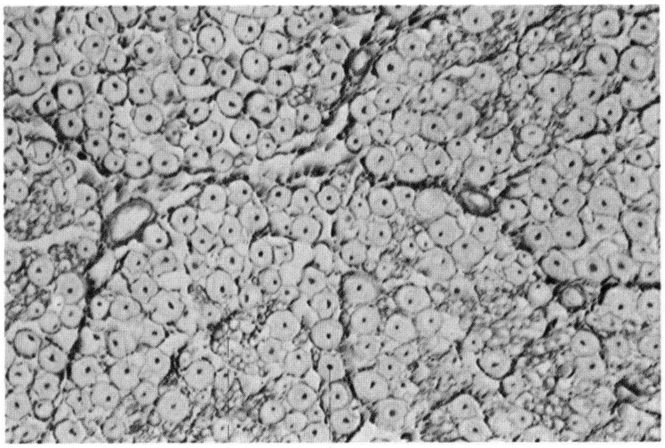

Fig. 174.

Fig. 175. Cross section of a myelinated peripheral nerve (human ischiadic nerve) whose shrunken axons appear as dark-violet or black spots, surrounded by a faint yellowish wrapping, the myelin sheath (cf. Fig. 177). Note the groups of small nerve fibers, either poor in myelin or completely devoid of it (unmyelinated), interspersed between the thick fibers with an elaborate myelin sheath (cf. Fig. 177). The Schwann cell nuclei cannot be identified, as no counterstain was performed. Staining: Picric acid and indigocarmine. Magnification 240×.

Small myelinated fibers   Axon

Fig. 175.

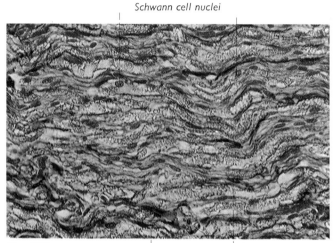

Schwann cell nuclei

Fig. 176.

Nuclei of endoneurial fibroblasts

Fig. 176. Longitudinal section through dorsal root of human spinal cord. In routine histologic preparations much of the myelin sheath is dissolved as a result of the use of lipid solvents, leaving behind a proteinaceous residue called neurokeratin. While the large elliptical nuclei belong to Schwann cells, the flat elongated ones belong to the fibroblasts of the endoneurium. Mallory-azan staining. Magnification 240×.

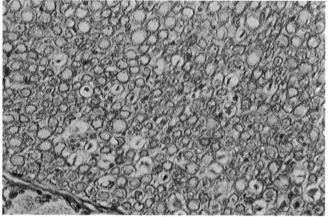

Fig. 177.

Axon

Fig. 177. Cross section through feline spinal nerve whose axons of various sizes give a finely granulated appearance of their myelin sheath (= neurokeratin), due to the extraction of lipids (cf. Fig. 176). Here and there the axon cylinders are shrunken and condensed into a centrally located deeply redstaining mass. Mallory-azan staining. Magnification 240×.

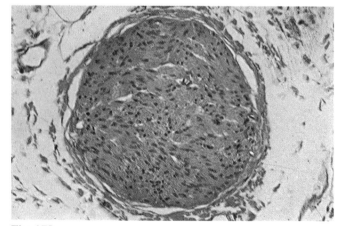

Fig. 178.

Fig. 178. Cross section through a small unmyelinated nerve from the vascular bundle of human spleen. Note that subdivisions by connective tissue septa are lacking in this case and that the axons are cut in all planes due to their twisting course. Another characteristic of unmyelinated nerves is the high amount of nuclei, i. e., cells. H.E. staining. Magnification 240×.

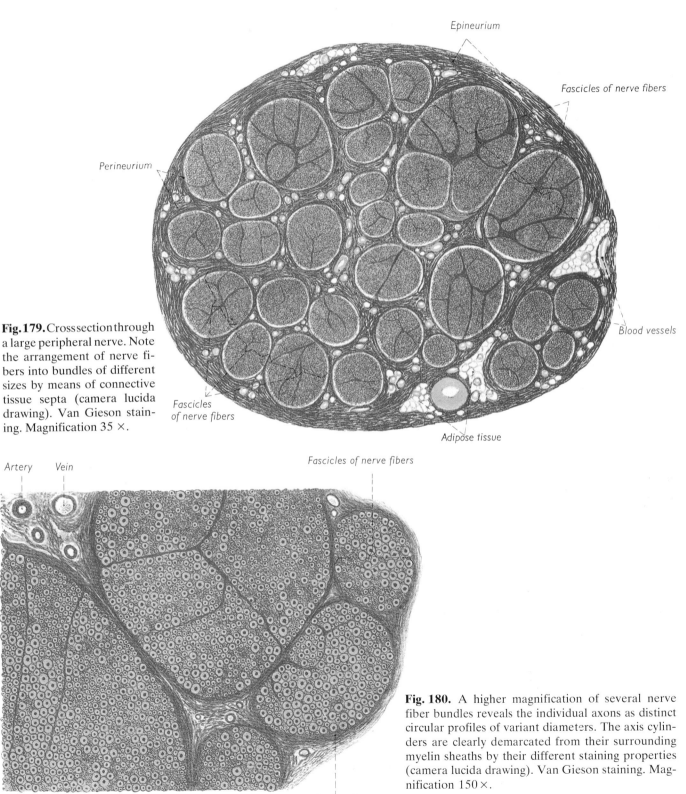

Epineurium

Fascicles of nerve fibers

Perineurium

Blood vessels

Fig. 179. Cross section through a large peripheral nerve. Note the arrangement of nerve fibers into bundles of different sizes by means of connective tissue septa (camera lucida drawing). Van Gieson staining. Magnification 35 ×.

Fascicles of nerve fibers

Adipose tissue

Artery   Vein

Fascicles of nerve fibers

Fig. 180. A higher magnification of several nerve fiber bundles reveals the individual axons as distinct circular profiles of variant diameters. The axis cylinders are clearly demarcated from their surrounding myelin sheaths by their different staining properties (camera lucida drawing). Van Gieson staining. Magnification 150 ×.

Perineurium

83

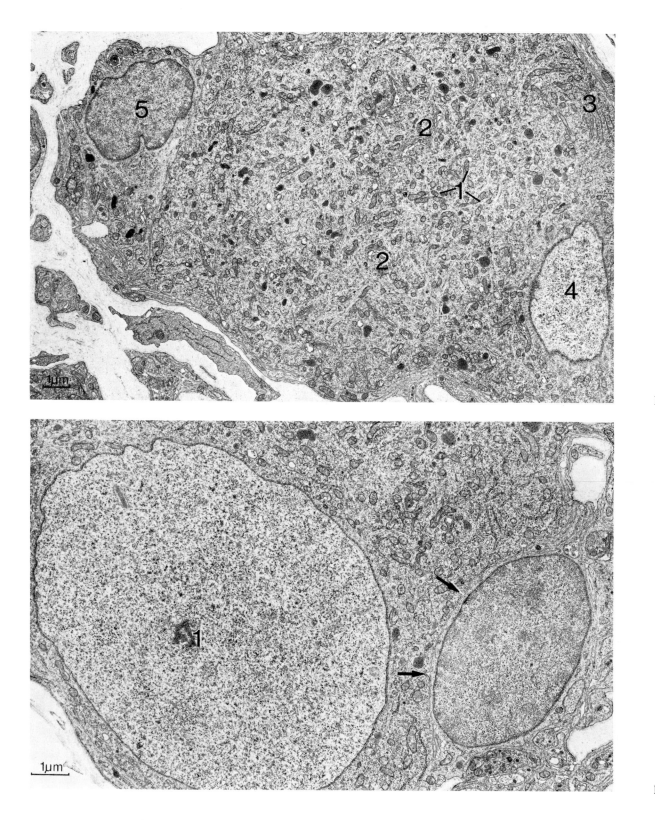

Fig. 181.

Fig. 182.

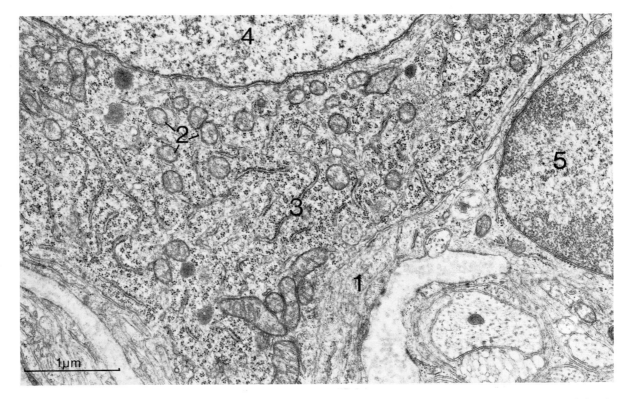

Fig. 183. Peripheral portions of an autonomous ganglion cell (same specimen as in the preceding figures), showing the origin of a dendrite with several satellite cell processes (1) closely attached to its surface. The nerve cell cytoplasm contains numerous mitochondria (2) together with many cisternae of a well-developed rough ER (3) and large amounts of free ribosomes. 4 = Nucleus of the nerve cell; 5 = Nucleus of a satellite cell. Magnification 24,500×.

◀

Fig. 181. Low-power view of a nerve cell in an autonomous ganglion from the feline pancreas. The perikaryon contains numerous but small mitochondria (1), several Golgi complexes (2) and a well-developed rough endoplasmic reticulum that here and there forms regular stacks of parallel cisternae (3). Only a small portion of the nucleus (4) positioned at the cell periphery are visible in this micrograph. 5 = Nucleus of a satellite cell. Magnification 7,000 ×.

Fig. 182. Another neuron from the same ganglion illustrates the large vesicular nucleus (1) together with its nucleolus and the extremely narrow space (→) intervening between nerve and satellite cell. Magnification 10,000×.

Nerve fibers, myelinated – Electron microscopy

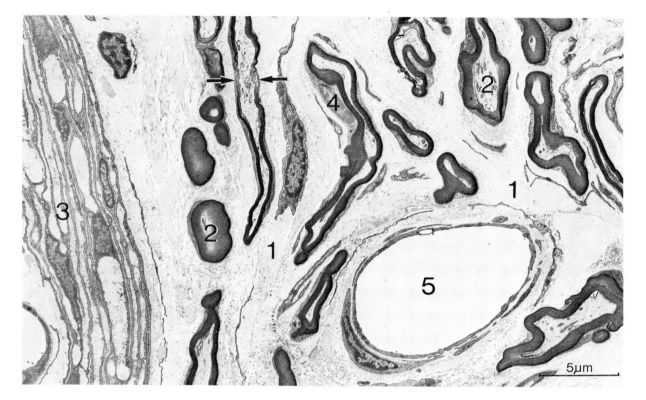

Fig. 184. Peripheral parts of a myelinated rabbit nerve with the axons (2) loosely arranged in an abundance of connective tissue (1 = collagenous fibrils) and cut in variant planes of section. The nerve is separated from its environment by a multilayered cellular sheath (3) composed of fibroblasts and their processes. The arrow points to a node of Ranvier. 4 = Nucleus of a Schwann cell; 5 = Postcapillary venule. Magnification 4,500×.

Fig. 185. Longitudunal section through an axon (rabbit) with myelin sheath (1) and (2) sheath of Schwann (= neurilemmal sheath). Note the intracytoplasmic bundle of filaments (3) in an adjacent fibroblast. 4 = Mitochondrium. Magnification 30,000×.

Fig. 186. a) Close-up of the preceding micrograph disclosing the characteristic layered appearance of the myelin sheath with a periodicity of 12 nm. The inner aspect of the myelin sheath is separated from the axoplasm (2) by the axolemma (1), while its outer surface is covered by the Schwann cell (3). Magnification 80,000×.
b) Longitudinal section through an unmyelinated nerve fiber with prominent neurofilaments (1) and neurotubules (2). Compare with Fig. 189. Magnification 54,000×.

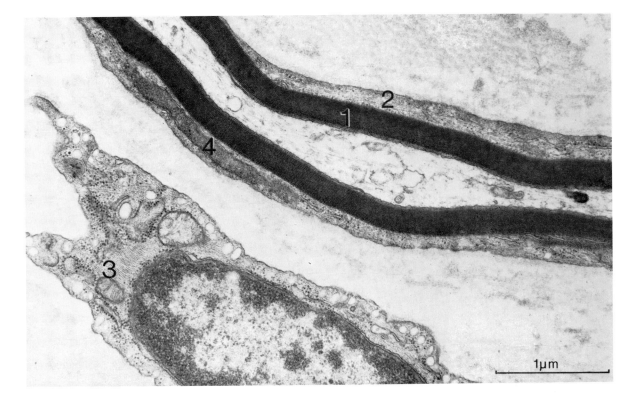

Fig. 185.

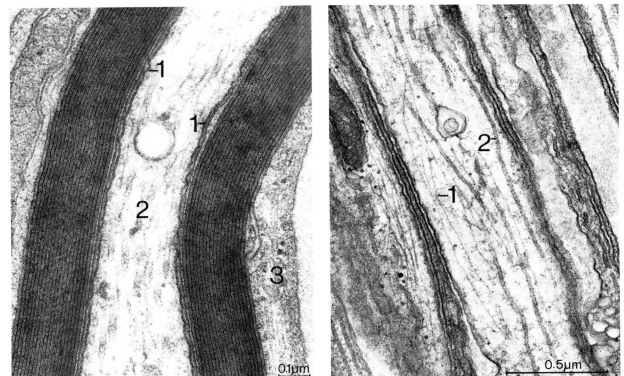

Fig. 186.

a

b

Nerve fibers, unmyelinated – Electron microscopy

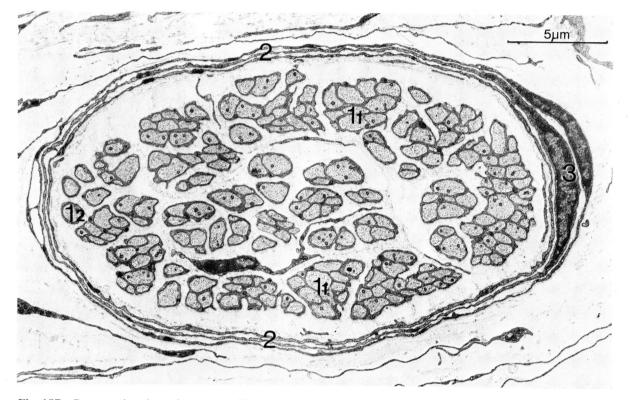

Fig. 187. Cross section through an unmyelinated autonomous rat nerve completely enclosed and thereby separated from the surrounding connective tissue by a multilayered sheath (2), the perithelium or perineural epithelium, which is composed of fibroblasts and their veil-like cytoplasmic processes. The darker staining cytoplasm of the Schwann cells enwraps the more translucent axons either in groups (1_1) varying in size and number, or it encircles merely a single axon (1_2) which all give a finely dotted appearance at this low magnification. 3 = Nucleus of a fibroblast. Magnification 5,000×.

Fig. 188. Part of a similar, yet not identical autonomous rat nerve to the one in the preceding micrograph showing either single (1_2) or multiple axons (1_1) enclosed in an often complex fashion ($\rightarrow$) by the thin cytoplasmic lamellae of a Schwann cell. 2 = Mitochondria. Magnification 24,000×.

Fig. 189. High-power electron micrograph of two axons from the preceding figure clearly exhibiting bundles of loosely associated filaments (1) and the coarser microtubules (2) that occur separately. Both these structures taken together represent the submicroscopic material visualized as neurofibrils in light microscopy following silver impregnation techniques. Note that the free edges (*) of delicate Schwann cell lamellae overlap each other over a longer distance. Magnification 53,000×.

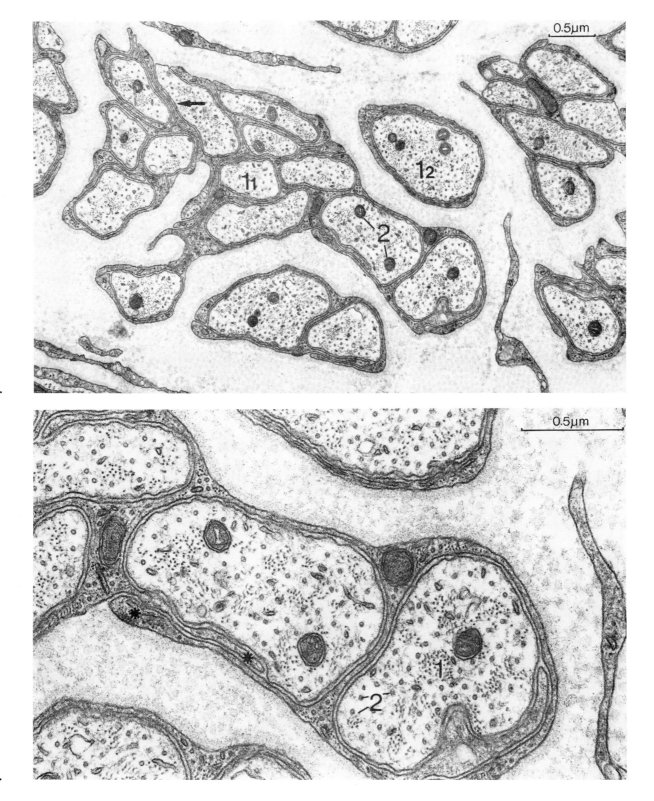

Fig. 188.

Fig. 189.

Nervous tissue – Neuroglia

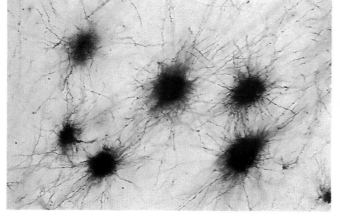

Fig. 190.

The demonstration of the different types of glia cells in the central nervous system can only be achieved by the involvement of special and rather "tricky" staining procedures. Therefore, in many histological courses only a restricted number of such rather precious preparations are available. (The specimes for Fig. 190, 191, 194 and 195 were kindly supplied by Prof. Dr. G. Kersting, Bonn, FRG.)

Fig. 190. Fibrous astrocytes from human cerebral white matter as seen with the Golgi method. The cell body often appears enlarged in such preparations due to the large amount of silver granules accumulating on the cell surfaces, thus even obscuring the nucleus. Clearly visible, however, are the numerous long and infrequently branched processes projecting in all directions from the perikarya, after which these cells were designated as "fibrous." Staining: Golgi's chrome-silver method. Magnification 240×.

Oligodendrocyte

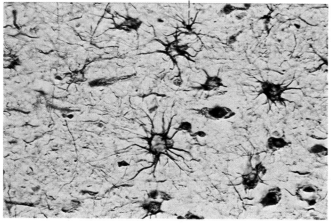

Fig. 191.

Fig. 191. Several astrocytes from human cerebral gray matter displaying a large cell body with short but highly branched processes and hence termed "protoplasmic astrocytes." In the upper part of the micrograph a smaller oligodendrocyte may be seen with fewer and more delicate processes. Staining: Bielschowsky's method. Magnification 380×.

———Astrocytes———

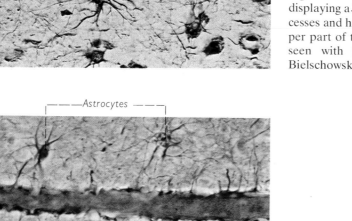
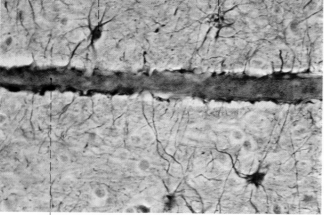

Small vein

Fig. 192.

Fig. 192. Fibrous astrocytes from human cerebral cortex that enclose the smallest blood vessels with the foot-shaped expansions of their processes, thus forming a perivascular glial sheath. Staining: Held's method. Magnification 380×.

Astrocytic cell body

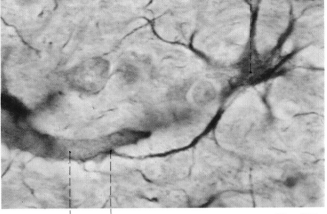

Capillary   Astrocytic process   **Fig. 193.**

Fig. 193. The higher magnification clearly reveals the extension of a fibrous astrocyte closely attached to a capillary (from human cerebral cortex). Staining: Held's method. Magnification 960 ×.

Oligodendrocyte

Fig. 194. Oligodendrocytes from human cerebral cortex. Their entire cell body is smaller than that of the astrocyte an hence nearly completely filled by the nucleus, as is true of the lymphocytes. Therefore only the nucleus can be seen in routine preparations by which the cells are difficult to identify. The oligodendrocytes are frequently found directly adjacent to nerve cell bodies, as seen in this micrograph, and then classified as perineural or satellite cells. Staining: Cajal's method. Magnification 380 ×.

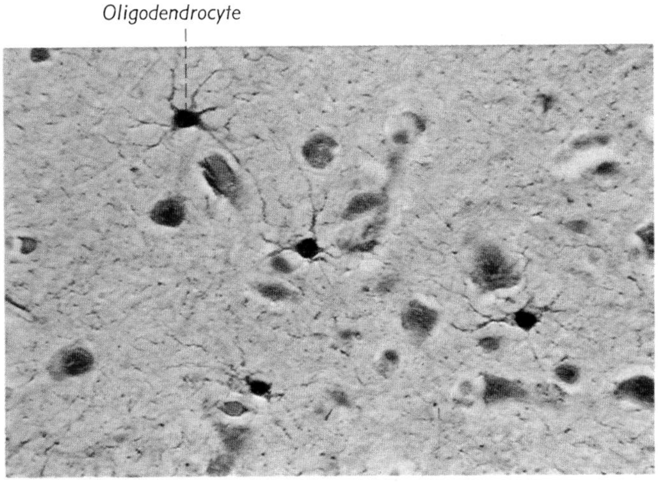

Fig. 194.

Fig. 195. Microglial cells from human cerebral cortex. These cells are small and give off only a few delicate and tortuous processes with spines. They are believed to be capable of ameboid movements and phagocytosis and therefore play a role in the removal of cellular debris in a variety of pathological conditions, e.g., following an apoplectic stroke. Staining: Hortega's method. Magnification 380 ×.

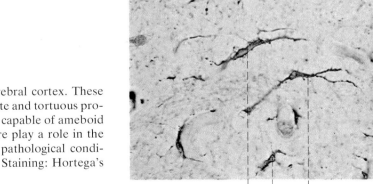

Fig. 195.   Microglial cells

Microscopic Anatomy

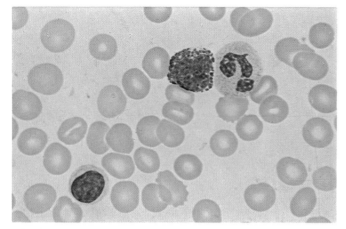

Fig. 196.

One of the most common routine tests in clinical medicine is the differential blood count, i.e., the determination of the percentages of the different types of white blood corpuscles (leucocytes) in dried blood smears. But as the special staining procedures by which the leucocytes can be subdivided into different types according to varying structural and staining properties need skilled and experienced technical assistants to achieve perfect results, one often is confronted with poor quality in such preparations.

The staining in the following micrographs is uniformly May-Grünwald.

Fig. 196. Three different types of leucocytes. In the upper part of the micrograph a cell filled with large basophilic granules, and hence a "basophilic granulocyte," is lying beside a neutrophilic polymorphonuclear leucocyte. At the lower left a lymphocyte can be seen with its characteristic nucleocytoplasmic relationship being considerably in favor of the nucleus (large nucleus surrounded by a small rim of cytoplasm). Note the various sizes of the different types of leucocytes and compare with each other and with the erythrocytes, as this is one of the criteria essential for a correct classification. Magnification 960×.

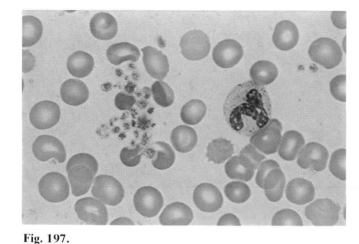

Fig. 197.

Fig. 197. Between the erythrocytes a cluster of blood platelets (= thrombocytes) can be seen, but their structural details cannot be identified at this low magnification. The neutrophilic granulocyte shows extremely fine granules corresponding to small pleomorphic lysosomes at the level of the electron microscope together with a rod-like, moderately lobed nucleus. Magnification 960×.

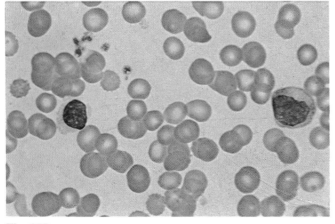

Fig. 198.

Fig. 198. While on the left side of the micrograph a "large" lymphocyte can be seen, a monocyte characterized by its large indented and often bean-shaped nucleus lies at the right side. At the upper left several blood platelets are visible. Magnification 750×.

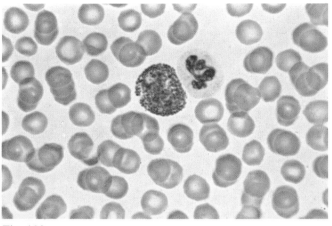

Fig. 199.

Drumstick

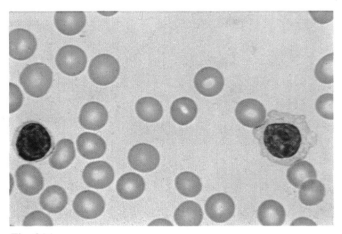

Fig. 200.

Fig. 199. Eosinophilic granulocyte with a bilobed nucleus, which often is one of the characteristic features of this type of leucocyte. If the granules are not stained as distinctly as in this case their number and size allow for their identification (cf. Fig. 197). At the level of the electron microscope the granules (= lysosomes) give a banded appearance characterized by a prominent crystal lattice in the center (cf. Fig. 204a). Magnification 960×.

Fig. 200. Neutrophilic granulocyte with a lobed nucleus showing a "drumstick" at its upper segment. This nuclear appendage represents the sex chromatin and is found with a frequency of 1 in every 36 neutrophils in women. But as this value varies and drumsticks are even found in the normal male with a maximum frequency of 1:1,000; 2,000 neutrophils must be evaluated in such a leucocyte test to allow for an exact chromosomal sex identification. Magnification 960×.

Fig. 201. This micrograph shows both a small (left side) and a large lymphocyte which can be clearly discriminated by their different nucleocytoplasmic relationships. While the small lymphocytes exhibit only a thin rim of cytoplasm, which sometimes is difficult to identify as such, the faintly staining cytoplasm of the younger but larger lymphocytes displays extremely fine azurophilic granules. Magnification 960×.

Fig. 202. Monocyte with a large indented nucleus that does not necessarily have to be bean-shaped, but never shows a circular outline with such regularity as the nuclei of the "large" lymphocytes. In its faintly staining basophilic cytoplasm, fine azurophilic granules can also be identified. Magnification 960×.

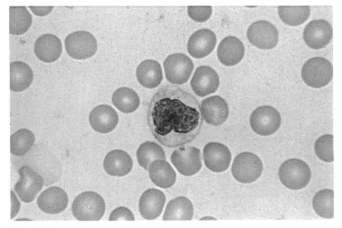

Fig. 201.

Fig. 202.

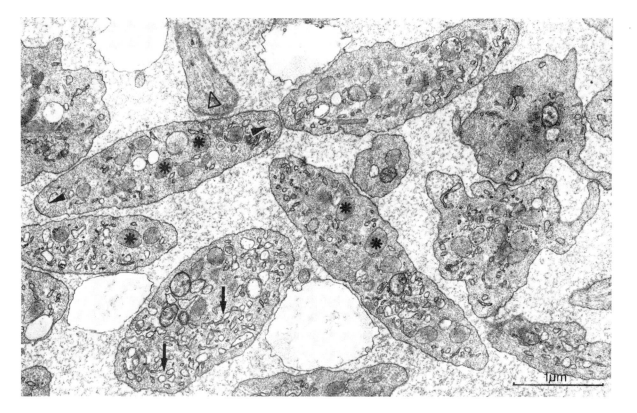

Fig. 203. Several human thrombocytes obtained from an experimentally prepared platelet-rich plasma. Together with membrane-bound granules of moderate electron density [α-granulomere (*)] vesicular and tubular profiles varying in outline [γ- granulomere ($\rightarrow$)] are visible. The circumferential bundle of microtubules positioned at the equatorial level of each platelet is seen in cross section ► at and in longitudinal section at ▷. Magnification 28,000 ×.

Fig. 204. Electron micrographs of various types of leucocytes. ►

a) The eosinophilic leucocytes (from rat intestinal lamina propria) are characterized by the large size and the shape of their specific granules that in addition possess a central discoid crystal. As in neutrophils, these granules are to be classified as lysosomes. Magnification 14,000 ×.

b) Neutrophilic leucocyte of a mouse with several nuclear lobes (1) that seem to be insulated because the interconnecting nuclear strands are beyond the plane of section. The specific granules (2) represent primary lysosomes that vary in size and electron density. Magnification 14,000 ×.

c) Basophilic leucocyte of a cat with remarkebly large granules (*) that show almost homogenous contents. The membrane that encloses these particles that are rich in heparin and histamine is seldom visible ($\rightarrow$). The nucleus is not cut in this section. Magnification 22,500 ×.

d) Lymphocyte from fresh human blood with numerous irregular processes along its surface. Note the circular outline of the nucleus and the nucleocytoplasmic relationship extremely in favor of the nucleus. Magnification 14,000 ×.

Fig. 204

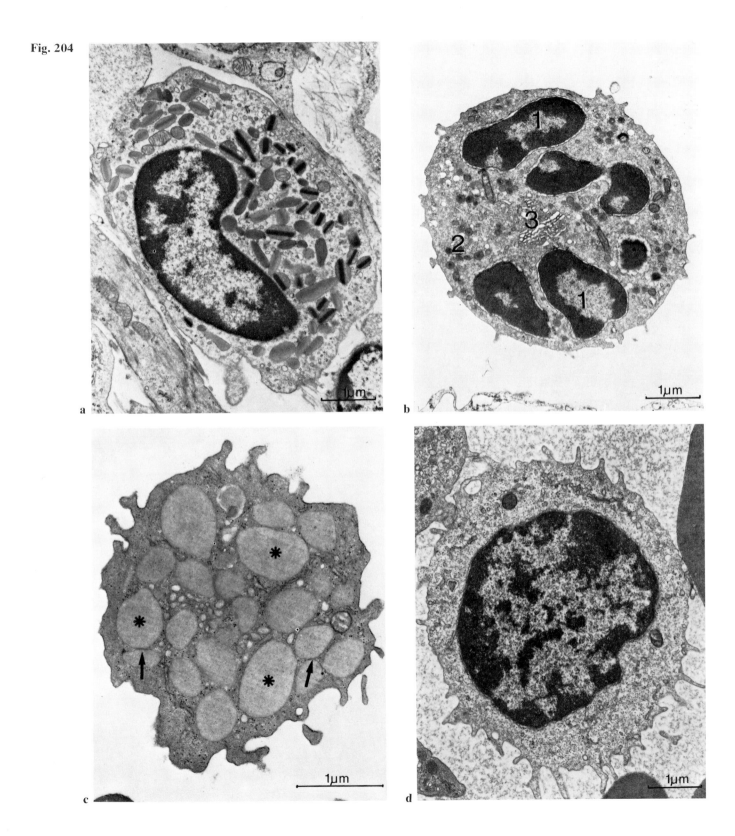

Red bone marrow and reticulocytes

Adipose cell

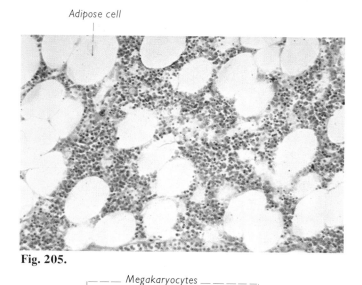

Fig. 205.

Megakaryocytes

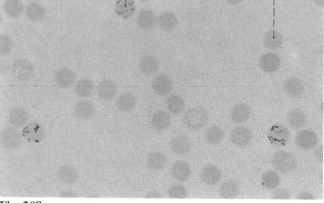

Fig. 207.

Normoblasts

Reticulocyte   Normocyte

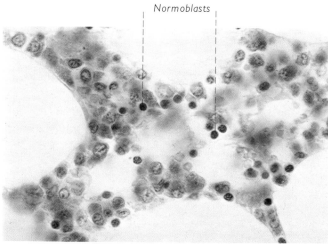

Fig. 206.

Normoblasts

Fig. 208.

Fig. 205. Red, hemopoietic bone marrow in situ (section through the spongiosa of a juvenile femur diaphysis) with numerous fat cells interspersed between its cellular strands, which consist of reticular connective tissue that is stuffed with innumerable cells belonging to the various developmental stages in erythro- and granulocytopoiesis. At the right an osseous trabecula is visible. H.E. staining. Magnification 95×.

Fig. 206. A higher magnification of the same specimen as shown in the preceding micrograph reveals various stages of the granulo- and erythropoiesis. Yet only the normoblasts can be definitely identified by their round and dense nuclei. H.E. staining. Magnification 600×.

Fig. 207. Three megakaryocytes from human bone marrow. These cells are not only characterized by their large size, but also by their apparent polynucleosis due to the variant and complicated arrangement of the many nuclear lobules. These cells give rise to the blood platelets that originate by fragmentation from the megakaryocytic pseudopodia. The dense round nuclei seen in this specimen belong to normoblasts. H.E. staining. Magnification 380×.

Fig. 208. Reticulocytes (do not confuse this term with reticulum cells) from the peripheral blood. The supravital staining with brilliant cresyl blue (this is performed by mixing fresh blood with the stain prior to preparing the smear) displays a fine granular network in these not fully matured erythrocytes, which is due to a precipitation and aggregation of ribosomes caused by the dye. An increase of reticulocytes (normally 12⁰/₀₀ of the erythrocytes) in the peripheral blood is an index for an increased rate of red cell formation in the bone marrow, e.g., following severe hemorrhages. Supravital staining with brilliant cresyl blue. Magnification 960×.

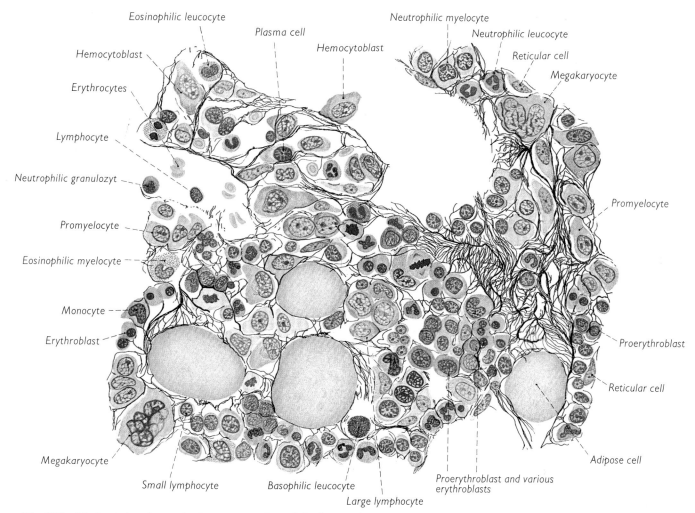

Eosinophilic leucocyte

Plasma cell

Hemocytoblast

Neutrophilic myelocyte

Neutrophilic leucocyte

Hemocytoblast

Reticular cell

Erythrocytes

Megakaryocyte

Lymphocyte

Neutrophilic granulozyt

Promyelocyte

Promyelocyte

Eosinophilic myelocyte

Monocyte

Erythroblast

Proerythroblast

Reticular cell

Megakaryocyte

Adipose cell

Small lymphocyte

Basophilic leucocyte

Proerythroblast and various erythroblasts

Large lymphocyte

Fig. 209. Moderately schematized representation of the human bone marrow and its supporting network of reticular fibers (from Patzelt: Histologie, 3rd ed., 1948). Staining: H.E. combined with silver impregnation. Magnification approx 1,200×.

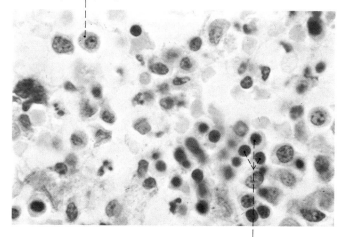

Eosinophilic myelocyte

Fig. 210. Original micrograph of human bone marrow that is at least comparable with the preceding drawing. While the normoblasts are distinct due to their round dense nuclei and the myeloblasts stand out by their specific granules, the majority of cells cannot be identified. Giemsa staining. Magnification 960×.

Normoblasts

The lymphatic organs may be subdivided into lymphoreticular and lymphoepithelial organs. The latter are mainly represented by the three tonsils, all of which show a combination of an epithelial surface with a supporting lymphatic tissue. The identification of the lymphoepithelial organs is based on 1) the different epithelia (only the pharyngeal tonsil shows a respiratory epithelium), 2) the size of the entire organs that are mostly cut as a whole (the palatine tonsil is much larger than the two others) and 3) the components of the surrounding tissues (only the lingual tonsils show larger amounts of glandular tissue).

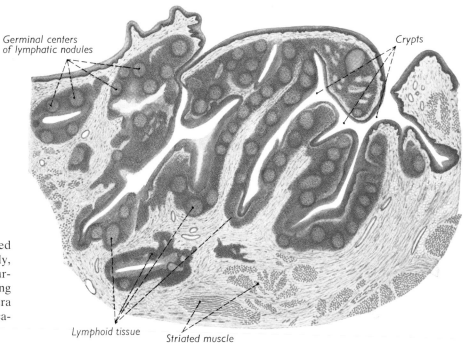

Fig. 211. Palatine tonsil whose stratified squamous epithelium invaginates deeply, thus forming branching crypts that are surrounded by lymphatic tissue containing numerous secondary nodules (camera lucida drawing). H.E. staining. Magnification 8×.

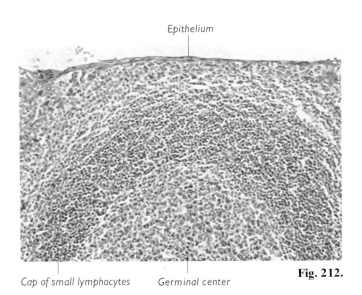

Fig. 212.

Fig. 212. Part of a crypt from a human palatine tonsil. The stratified nonkeratinized squamous epithelium is reduced to two or three thin layers while the rest of it is infiltrated by large numbers of lymphocytes and thereby is transformed into a loose cellular network, an epithelial reticulum. Beyond this, parts of a secondary lymphatic nodule with its germinal center and cap of small lymphocytes are visible. H.E. staining. Magnification 150×.

Crypt

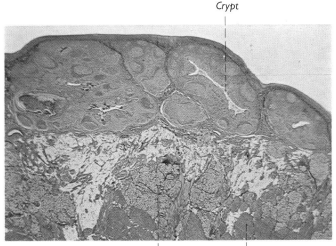

Fig. 213.

Mucous glands   Skeletal muscle fibers

Germinal center   Epithelial remnants

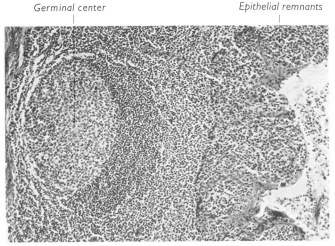

Fig. 214.

Fig. 213. The lingual tonsil, with less branching and shorter crypts than the palatine tonsils, are surrounded not only by the skeletal muscle of the tongue (the occurrence of skeletal muscle near a tonsil is in itself no definite criterion for the identification; cf. the foregoing micrograph), but by numerous, predominantly mucous glands. Mallory-azan staining. Magnification 12×.

Fig. 214. Close-up of the lower right portion of the crypt indicated in the preceding micrograph by a leader line. The epithelium of the crypt is almost entirely transformed into a lympho-epithelial network with a few patches of epithelial remnants on top of it. Immediately above the connective tissue a secondary lymphatic nodule with a germinal center and its crescent-shaped cellular cap is illustrated. Mallory-azan staining. Magnification 96×.

Fig. 215. The pharyngeal tonsil is not only the smallest of all these organs, but the only one covered by a ciliated pseudo-stratified columnar epithelium. Well-preserved and healthy specimens of this tonsil are difficult to obtain, as it is well-developed only in young individuals and hence it is but rarely found in routine histology courses. Mallory-azan staining. Magnification 13×.

Fig. 216. Higher magnification of the right half of the cone-shaped area of the pharyngeal tonsil from the preceding micrograph. Now the respiratory epithelium can be clearly identified, yet it appears remarkably disintegrated at the sites of its sinuate indendations. Mallory-azan staining. Magnification 96×.

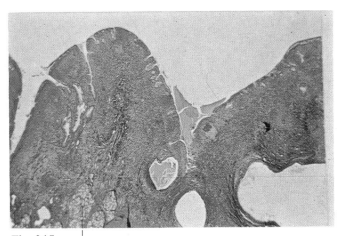

Fig. 215.   Seromucous glands

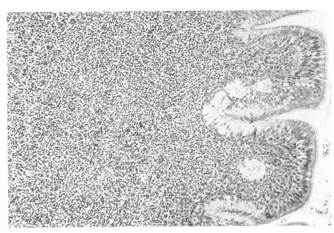

Fig. 216.

Lymphatic organs – Spleen

A first identification of the lymphoreticular organs consisting of the lymph nodes, the spleen and the thymus is already possible by the naked eye eventually assisted by a magnifying glass, as two of these – the lymph nodes and the thymus – show a subdivision into a centrally located medulla surrounded by an outer cortex. In addition only the lymph nodes possess a "marginal sinus" running immediately below the capsule. In contrast the spleen neither displays a medulla-cortical organization nor a subcapsular sinus, but contains many lymphatic nodules surrounding small arteries, which are known as Malpighian corpuscles. A definite characteristic of the thymus is the medullary bodies (Hassall's corpuscles).

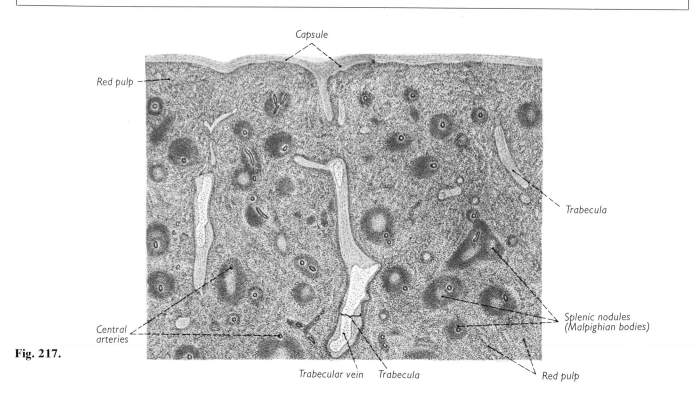

Fig. 217.

Capsule

Red pulp

Trabecula

Central arteries

Splenic nodules (Malpighian bodies)

Trabecular vein   Trabecula

Red pulp

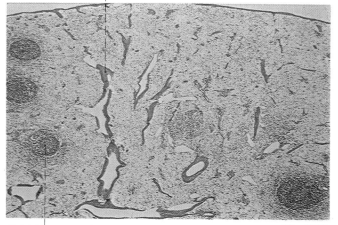

Trabecula with trabecular vein

Malpighian corpuscles   **Fig. 218.**

Fig. 217. General view of a subcapsular area of human spleen showing several Malpighian corpuscles. These consist of lymphatic tissue with occasional germinal centers forming elongated cylindrical sheaths around certain divisions of the arteries, and as a whole represent the white pulp of the spleen. The arteries in these splenic corpuscles are called central arteries, although they are eccentrically located within their lymphatic sheaths. The fibrous capsule is continuous with the trabeculae, carrying the larger blood vessels and traversing the entire organ, thus forming a coarse connective tissue framework. In some species, e.g., cats, the capsule contains numerous smooth muscle cells (camera lucida drawing). H.E. staining. Magnification 22×.

Fig. 218. Feline spleen that prior to fixation has been profusely perfused via its artery to remove most of the blood it contains. Hence its reticular connective tissue and the finer branches of the vascular tree usually obscured by the abundance of erythrocytes are shown to a better advantage. The Malpighian corpuscles remain unaltered by this procedure. H.E. staining. Magnification 24×.

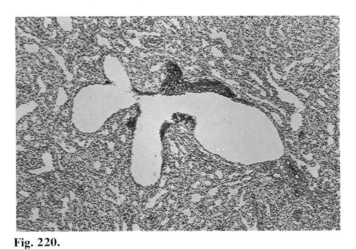

Fig. 219. An artery traversing the red pulp of a rhesus monkey and reaching with both its branches into the lymphatic tissue sheaths contributed by the white pulp, thus becoming the central arteries of the Malpighian corpuscles. H.E. staining. Magnification 95×.

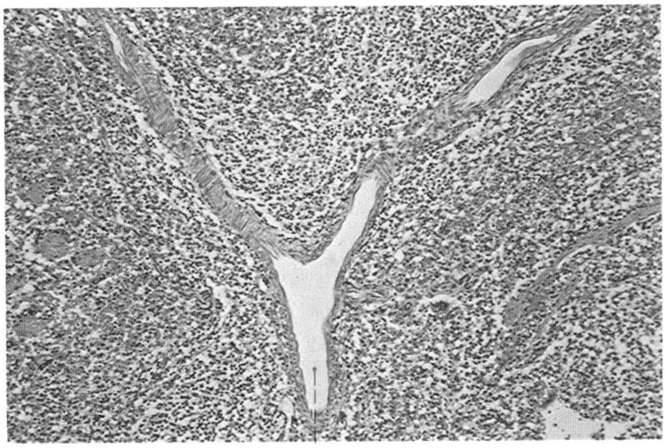

Fig. 219.   *Pulp artery*

Fig. 220. Low-power micrograph of the red pulp of human spleen showing a trabecular vein with some of its larger tributaries. Mallory-azan staining. Magnification 60×.

Fig. 220.

Lumen of venous sinus   *Reticular fibers*

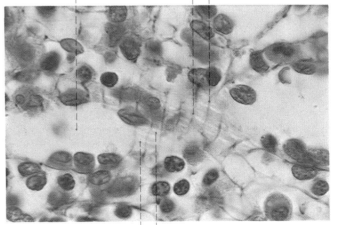

Fig. 221. At higher magnification, the venous sinuses display structural details of their walls, particularly when cut tangentially, thus allowing for a surface view (center of this micrograph). They consist of elongated, longitudinally arranged cells (= lining reticular or littoral cells) that stain faintly and are encircled by relatively coarse reticular fibers. In transverse sections the endothelial nuclei can be seen bulging into the lumen. Mallory-azan staining. Magnification 960×.

Fig. 221.   *Endothelial cell of venous sinus (long. sec.)*

103

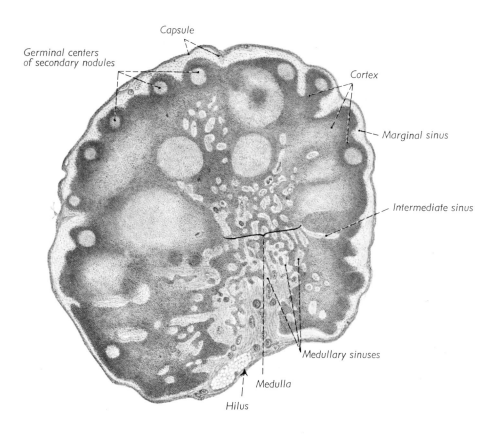

Fig. 222. Section through an entire human lymph node with an extremely broad subcapsular marginal sinus but an indistinct organization into medulla and cortex. Germinal centers or secondary nodules are almost exclusively found in the primary nodules of the cortex, which also contains many more cells and hence is stained deeper than the medulla (camera lucida drawing). H.E. staining. Magnification 18×.

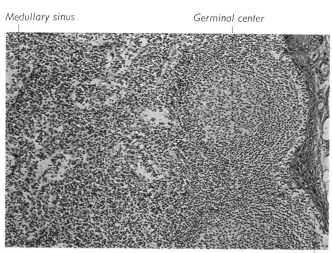

Fig. 223.

Fig. 223. Part of a human lymph node cortex at a higher magnification. The narrow marginal sinus is filled with an abundance of lymphocytes and defined by a fibrous capsule in which numerous medium-sized blood vessels and larger lymphatics may be seen. The marginal sinus may be difficult to identify as such if completely stuffed with cells, as in inflammatory reactions, and it can be simulated in the spleen by a subcapsular cleft caused by shrinkage (caution). In the cortex below the marginal sinus, a primary follicle with a germinal center is clearly visible. The medulla is characterized by its many and rather broad lymphatic sinuses (= medullary sinuses) with fewer and more loosely arranged cells. Mallory-azan staining. Magnification 95×.

The thymus is characteristically organized into lobules, each of which is subdivided into a medulla and a cortex. Moreover, the thymus is devoid of both a marginal sinus and secondary nodules, but shows Hassall's corpuscles in the medulla.

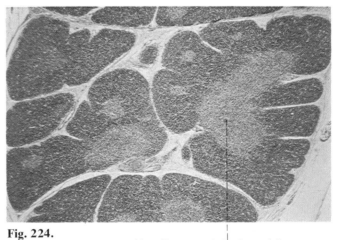

Fig. 224. Well-developed thymus of a human fetus showing a prominent lobulation together with a clear-cut division into medulla and cortex. The latter is more deeply stained due to its abundance of cells. Hematoxylin-chromotrop staining. Magnification 24×.

Fig. 224.

Hassall's corpuscle in the medulla

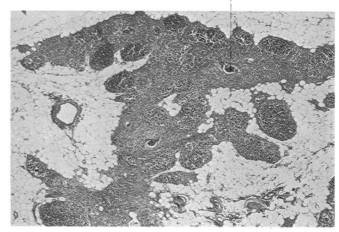

Fig. 225. In the thymus of adults the lobulation disappears almost completely due to an involution of the cortex ("age involution"). In the persisting medullary cords, conspicuously large, often cyst-like, Hassall's corpuscles can be found filled with a lumpy and disintegrating material. Hematoxylin-chromotrop staining. Magnification 24×.

Fig. 225.

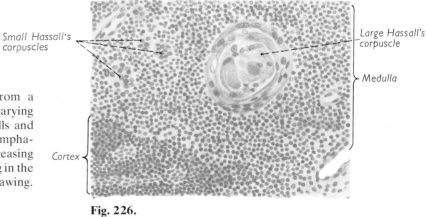

Small Hassall's corpuscles

Large Hassall's corpuscle

Medulla

Fig. 226. Well-developed Hassall's corpuscle from a child's thymus. These bodies are composed of a varying number of concentrically arranged medullary cells and represent the most definite characteristic of this lymphatic organ. With progressive age they show an increasing degeneration of their central parts, finally resulting in the formation of cysts (cf. Fig. 225). Camera lucida drawing. Alum-carmine staining. Magnification 230×.

Cortex

Fig. 226.

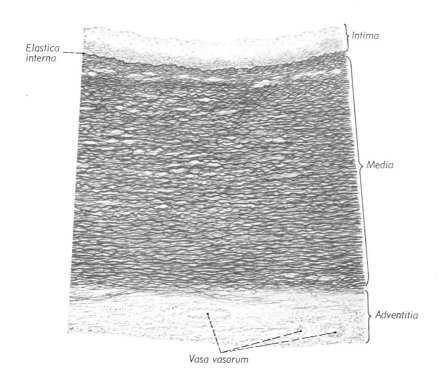

Elastica
interna

Intima

Media

Adventitia

Vasa vasorum

Fig. 227. Segment of a cross-sectioned human aorta with an elastic fiber stain showing the three main tunics characteristic of each artery: 1) intima, 2) media and 3) the adventitia composed of connective tissue. When compared, the internal and external elastic laminae are less pronounced in the arteries of the elastic type to which belong the pulmonary artery and the aorta with its primary branches. The smooth muscle cells in the media are unstained and hence invisible (camera lucida drawing). Orcein staining. Magnification 60×.

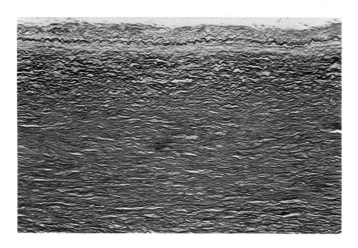

Fig. 228. Part of the human aortic wall with the intima (at top) and inner half of the media in which not only the great masses of elastic tissue, but also the smooth muscle cells are stained. Resorcin-fuchsin staining combined with Goldner staining. Magnification 95×.

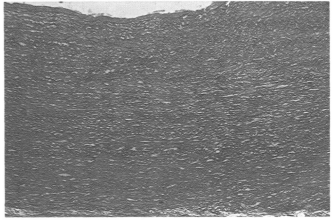

Fig. 229.

Fig. 229. Low-power view of a cross-sectioned human descending aorta that is often difficult to identify correctly in H. and E. stain because of an indistinct separation of the different tunics. In addition, the expected endothelial layer which lines the vessel often is not present due to post mortem changes, and the structural elements of the media (smooth muscle cells and elastic fibers) may only be identifiable at higher magnifications. Therefore, such preparations of the aorta are often mistaken as an "elastic ligament" or simply stated as "smooth muscle tissue." H.E. staining. Magnification 38×.

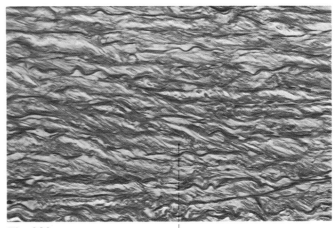

Fig. 230.

Smooth muscle cell

Fig. 230. Aortic media at a higher magnification with a combined stain for cells and elastic fibers to demonstrate the close interrelationships between the muscle cells and the numerous elastic membranes whose tensile strength is modulated by the action of the former. Staining: Resorcin-fuchsin and azocarmine-naphthol green. Magnification 240×.

Connective tissue beyond the intima   Smooth muscle within the intima

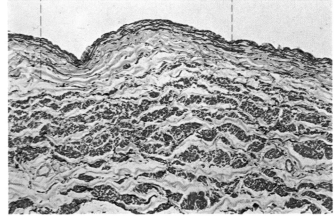

Fig. 231. Segment of the entire wall of a cross-sectioned human inferior vena cava. When compared with the aorta, the elements of the media are more loosely arranged and separated from the intima by a broad connective tissue layer (subintimal connective tissue). Immediately below its endothelium the intima contains small strands of smooth muscle cells (staining bright red). Staining: Resorcin-fuchsin and azocarmine-naphthol green. Magnification 95×.

Fig. 231.

Blood vascular system – Muscular arteries and veins

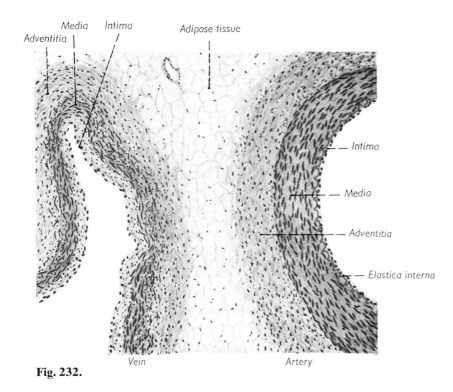

Adventitia Media Intima — Adipose tissue

— Intima

— Media

— Adventitia

— Elastica interna

Fig. 232. Vein   Artery

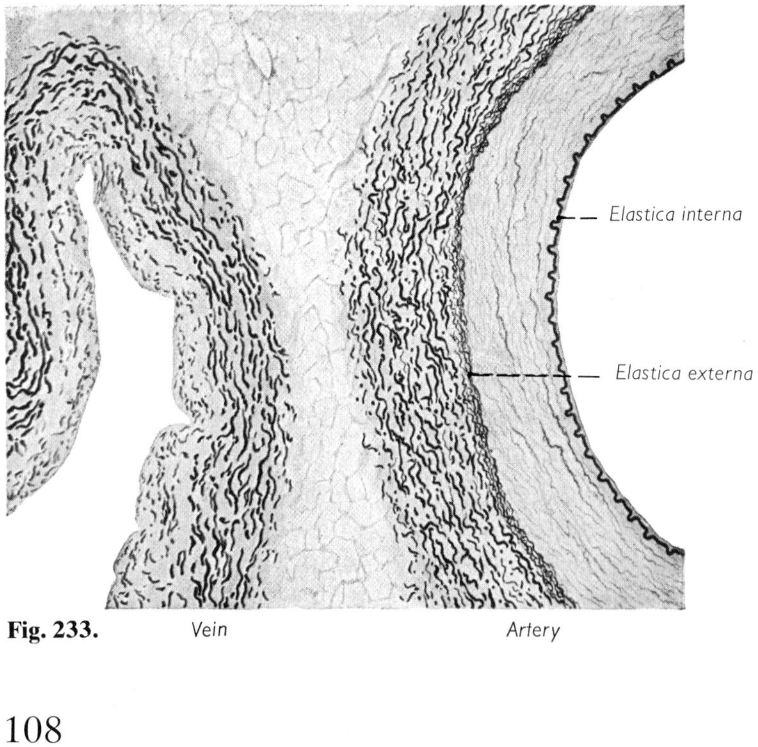

— Elastica interna

— Elastica externa

Fig. 233. Vein   Artery

Figs. 232 and **233.** Two cross sections of the same medium-sized muscular artery with its accompanying vein stained with hematoxylin-eosin and resorcin-fuchsin respectively to demonstrate the different compositions of their walls. A reliable criterion for the identification of these vessels is the structure of the media. In arteries it consists of closely apposed smooth muscle cells with only a small amount of connective tissue fibers interspersed, whereas the media of veins displays a looser arrangement of fewer muscle cells and a richer supply of collagenous fibers. The internal elastic lamina is generally more pronounced in typical arteries than in the corresponding veins (best seen with elastica stains), but may be found occasionally also in the latter (camera lucida drawing). H.E. staining (Fig. 232); Resorcin-fuchsing staining (Fig. 233). Magnification 65×.

Artery   Nerve

Skeletal muscle fibers   Artery

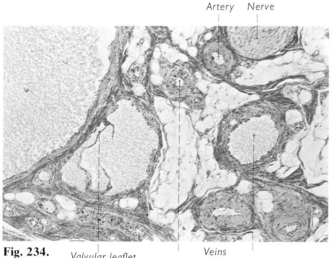

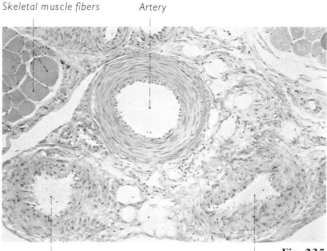

Fig. 234.   *Valvular leaflet in a lymphatic vessel*   *Veins*

Vein   *Vein*   **Fig. 235.**

Fig. 234. Low-power view of smaller blood and lymphatic vessels from human spermatic cord. In this particular location the veins have a media similar in width to that of the arteries, but they can be readily distinguished from the latter by both the lower number and the irregular distribution of the nuclei within their media (cf. the following micrograph). The vascular profile occupying the upper left corner of the micrograph and filled with a congealing content represents a larger lymphatic trunk to which a smaller tributary is closely attached that shows the leaflets of a valve in its lumen. H.E. staining. Magnification 60×.

Fig. 235. Small muscular artery together with its accompanying veins that in this special case show a media similar in width to that of the artery (from human spermatic cord). Note, however, the diversity in both the number and the arrangement of the nuclei in the arterial and in the venous media. In the upper left corner cross-sectioned skeletal muscle fibers (m. cremaster) can be seen. H.E. staining. Magnification 96×.

Fig. 236. Demonstration of capillaries (open circles) in the rat gastrocnemius muscle by vascular perfusion with a fixative (glutaraldehyde). In addition, a classification of muscle fiber types is possible in this specimen (cf. Fig. 145) as this semi-thin section (O,5μm) was treated with a histochemical method that stains mitochondria and their subsarcolemmal accumulations as brownish dots and crescentshaped areas respectively. The unusually high capillarization is due to the relatively large number of "red" fibers in this area. P-phenylendiamine staining. Magnification 240×.

Fig. 237. Cross section of a small, muscle-free lymphatic vessel from the human spermatic cord showing the two leaflets of a valve in its lumen. Mallory-azan staining. Magnification 240×.

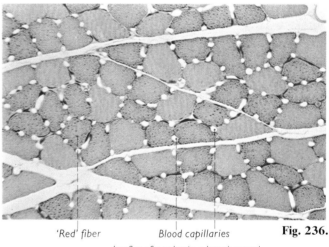

'Red' fiber   *Blood capillaries*   **Fig. 236.**

Leaflet of a valve in a lymph vessel

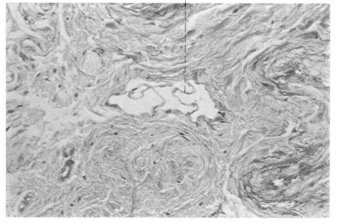

Fig. 237.

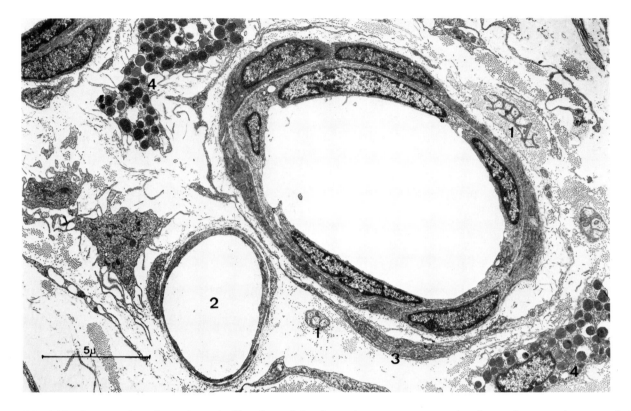

Fig. 238.

Fig. 238. Cross section of a moderately dilated arteriole of rat subcutaneous connective tissue whose continuous endothelium is surrounded by a single layer of closely apposed smooth muscle cells. Moreover, fine processes of fibroblasts (3) form an incomplete perivascular sheath. 1 = Bundles of unmyelinated nerves; 2 = Cross section of a capillary; 4 = Mast cells. Magnification 6,000×.

▶

Fig. 239. Transverse section of a capillary of the subcutaneous tissue of a rabbit's ear. Due to the continuity of its bicellular endothelial layer it belongs to the "continuous" type of capillaries and is almost completely enwrapped by the slender process of a pericyte (3). Within the endothelial cytoplasm groups of vesicles and vacuoles together with a few mitochondrial profiles can be seen; the pericyte shows not only its large nucleus and mitochondria, but a well-developed rough-surfaced endoplasmic reticulum and numerous free ribosomes as well. 1 = Interendothelial cleft; 2 = Basement membrane(▶). Magnification 13,500×.

Fig. 240. Cross section of a "fenestrated capillary" of the rat thyroid gland. The endothelium shows extremely attenuated areas perforated by regularly spaced circular pores (= fenestrae) with a diameter of approx. 50 nm and closed by a delicate membrane (= diaphragm). At (▶) the tangential sections of fenestrae clearly exhibit a central knob-like thickening of their diaphragms. 1 = Extremely dilated cisternae of the ergastoplasm of a follicular epithelial cell; 2 = Endothelial basal lamina; 3 = Endothelial nucleus. Magnification 22,500×.

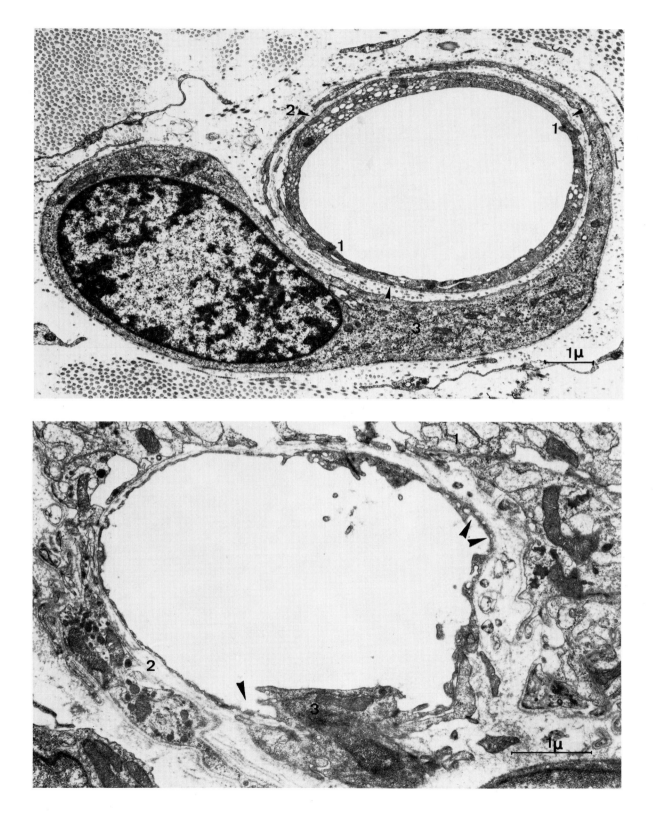

Fig. 239.

Fig. 240.

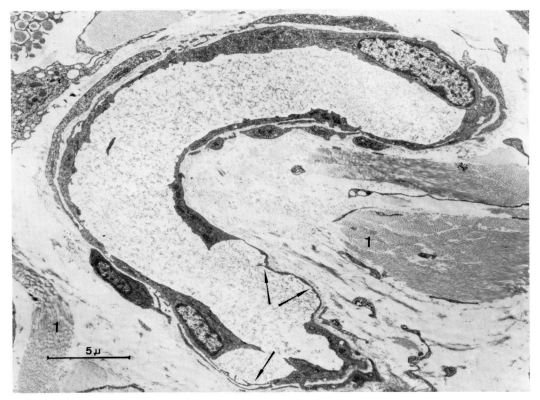

Fig. 241.

Fig. 241. Cross section of a postcapillary venule that is continuous with one of its tributaries, a venous capillary (subcutaneous connective tissue of the rat). The latter can be identified by its extremely flattened endothelium which still shows a few fenestrae ($\longrightarrow$), which are, however, undistinguishable due to the low magnification. The subendothelially located cytoplasmic profiles mainly belong to pericytes. 1 = Collagenous fibrils. Magnification 4,600×.

▶

Fig. 242. Medium-sized nonmuscular venule with erythrocytes and two blood platelets in its lumen (subcutaneous connective tissue of the rat). The few subendothelially located cells contain no filaments and hence are considered to be pericytes or possibly poorly differentiated muscle cells. Magnification 4,400×.

Fig. 243. Larger lymphatic vessel from the subcutis of the rat's paw, whose thin endothelium shows at ($\longrightarrow$) an open junction. This is a regular occurrence in the smaller lymphatics and serves as a preformed inflow channel for the interstitial fluid together with the macromolecules it contains. Unlike the veins a lymphatic vessel of the same size possesses 1) endothelial surfaces that are much more irregularly outlined, 2) no basal lamina or only fragments of it and 3) no cells immediately adjacent to the endothelial lining. The capillary – it is an arterial capillary segment – is stuffed with erythrocytes and platelets. In the cytoplasm of the pericyte (2) profiles of phagocytized erythrocytes are visible. 1 = Fibroblast. Magnification 5,600×.

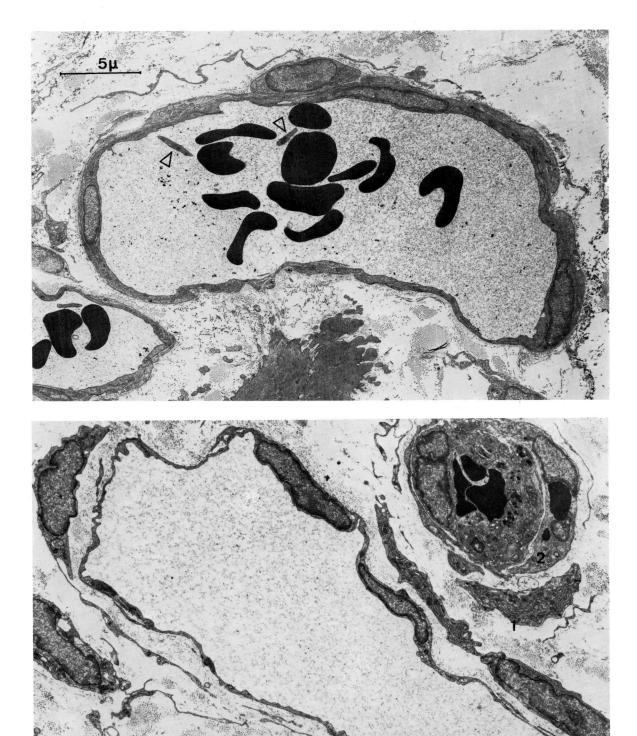

Fig. 242.

Fig. 243.

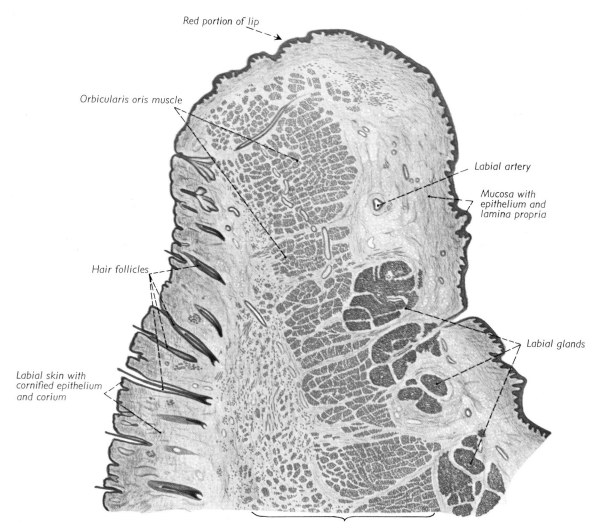

Red portion of lip

Orbicularis oris muscle

Labial artery

Mucosa with epithelium and lamina propria

Hair follicles

Labial glands

Labial skin with cornified epithelium and corium

Orbicularis oris muscle

Fig. 244. The lips belong to those areas that are characterized, besides by other features by a gradual change of their covering epithelium. In this sagittal section it becomes evident that a typical thin skin with a cornified epithelium, hair follicles and both sweat and sebaceous glands changes in the "red area" into a noncornified, stratified squamous epithelium devoid of any glands. This epithelium is continouous with a similar epithelium of the mucous membrane which, in its submucosa, constains numerous mixed glands. The central tissue core of the lips is mainly occupied by the striated fibers (cross-sectioned in this specimen) of the orbicularis oris muscle (camera lucida drawing). For detailed analysis for identification see Table 11. H. E. staining. Magnification 8×.

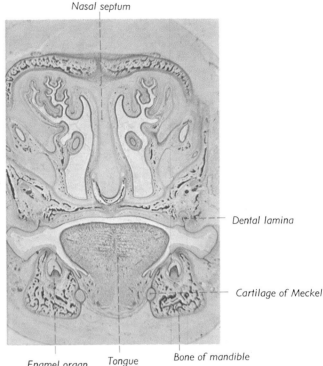

Nasal septum

Dental lamina

Cartilage of Meckel

Enamel organ   Tongue   Bone of mandible

Fig. 245. Frontal section through the snout of a porcine fetus. In the upper and lower jaw area (their osseous trabeculae stained a brilliant blue) tooth germs of different developmental stages can be seen. In the maxilla they are represented in the form of the early dental lamina (particularly prominent at the right side of the micrograph), while in the mandible they already consist of the epithelial enamel organ and the mesenchymal dental papilla. Mallory-azan staining. Magnification 9.5×.

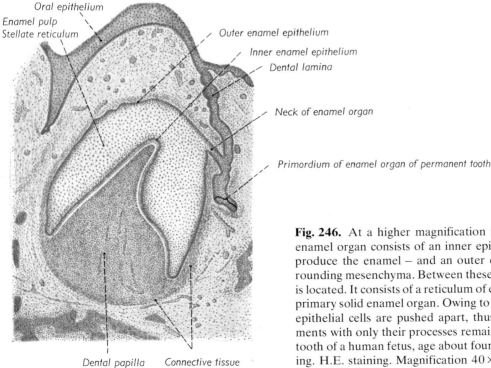

Oral epithelium
Enamel pulp
Stellate reticulum

Outer enamel epithelium

Inner enamel epithelium

Dental lamina

Neck of enamel organ

Primordium of enamel organ of permanent tooth

Dental papilla   Connective tissue

Fig. 246. At a higher magnification it can be shown that the bell-shaped enamel organ consists of an inner epithelium – the future ameloblasts that produce the enamel – and an outer enamel epithelium adjoining the surrounding mesenchyma. Between these two epithelial linings the enamel pulp is located. It consists of a reticulum of epithelial origin that develops from the primary solid enamel organ. Owing to an increase of the interstitial fluid, the epithelial cells are pushed apart, thus being transformed into stellate elements with only their processes remaining in contact (developing deciduous tooth of a human fetus, age about four to five months). Camera lucida drawing. H.E. staining. Magnification 40×.

115

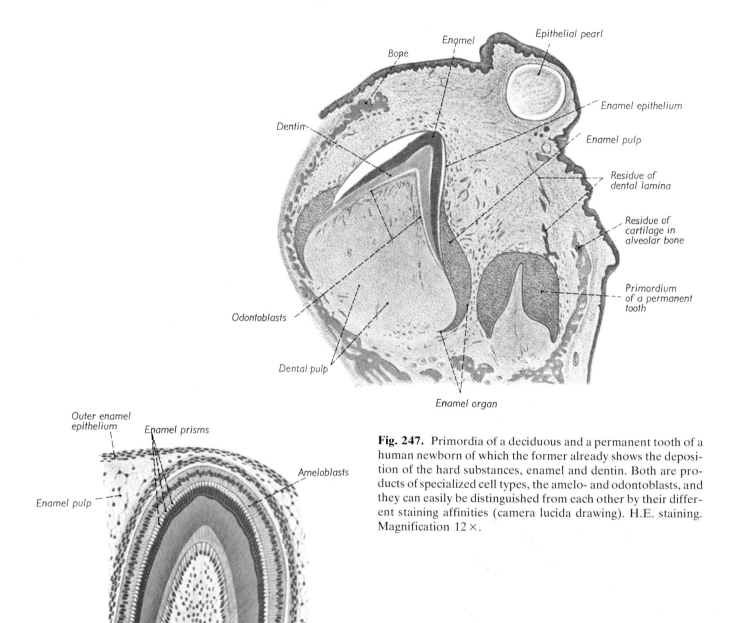

Bone

Enamel

Epithelial pearl

Dentin

Enamel epithelium

Enamel pulp

Residue of dental lamina

Residue of cartilage in alveolar bone

Odontoblasts

Primordium of a permanent tooth

Dental pulp

Enamel organ

Outer enamel epithelium

Enamel prisms

Ameloblasts

Enamel pulp

Dentin

Older-

Younger- (= predentin)

Blood vessels

Odontoblasts

Fig. 247. Primordia of a deciduous and a permanent tooth of a human newborn of which the former already shows the deposition of the hard substances, enamel and dentin. Both are products of specialized cell types, the amelo- and odontoblasts, and they can easily be distinguished from each other by their different staining affinities (camera lucida drawing). H.E. staining. Magnification 12×.

Fig. 248. Detail from the crown of human dental primordium (approx. age, 6 months) showing the first stages in the deposition of enamel and dentin. The odontoblasts originate from those mesenchymal cells of the dental papilla that are adjacent to the enamel organ. They first produce an uncalcified predentin (= dentinoid) in which cytoplasmic processes of the odontoblasts (= fiber of Tomes) survive and remain active. In contrast the ameloblasts elaborate their product as a cuticular secretion in the form of prisms that they push forward toward the dentin and thereby gradually withdraw from it (camera lucida drawing). H.E. staining. Magnification 165×.

Fig. 249. Complete longitudinal section through a cat's incisor in situ with its crown (that part that projects above the gingiva), neck (that portion where the enamel and the cement merge with each other) and root (the part located in an osseous socket or alveolus). In this specimen the enamel is invisible due to its removal by decalcification prior to sectioning (camera lucida drawing). H.E. staining. Magnification 18×.

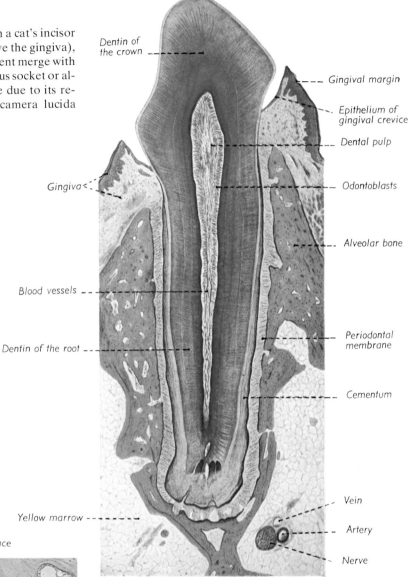

Dentin of the crown

Gingival margin

Epithelium of gingival crevice

Dental pulp

Odontoblasts

Gingiva

Alveolar bone

Blood vessels

Periodontal membrane

Dentin of the root

Cementum

Vein

Artery

Yellow marrow

Nerve

Connective tissue with blood vessels in the periodental space

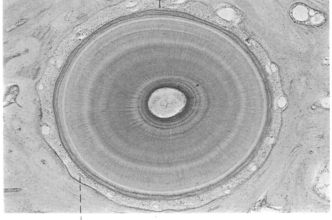

Alveolar bone

Fig. 250. Cross section through a feline's incisor root in situ. Its dentin, due to its stepwise calcification, shows a concentric layering. The growth lines between the older and the newly formed layers of dentin are known as contour lines of Owen. Hem. and picric acid staining. Magnification 38×.

Oral cavity – The tongue

Filiform papillae

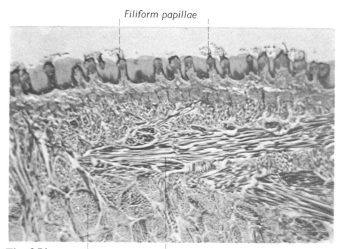

Fig. 251.

Bundles of skeletal muscle fibers

Fig. 251. Dorsum of human tongue with closely spaced filiform papillae. These consist of a connective tissue core that subdivides into secondary papillae whose covering epithelium tapers into threadlike (hence the name!) cornifications bent towards the pharynx. These papillae serve mechanical purposes. Hem. and azocarmine staining. Magnification 12 ×.

Serous glands

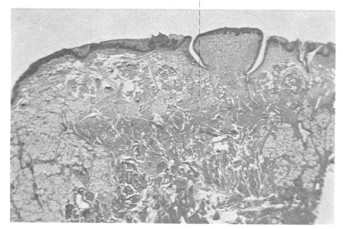

Fig. 252.

Fig. 252. The circumvallate papillae can be seen with the naked eye, being located at the junction of the lingual dorsum and root of the tongue. Numerous taste buds are interspersed in the epithelial walls of their trenches, into which open the ducts of the serous glands of von Ebner. Due to the low magnification of this specimen, the taste buds cannot be identified. H.E. staining. Magnification 12 ×.

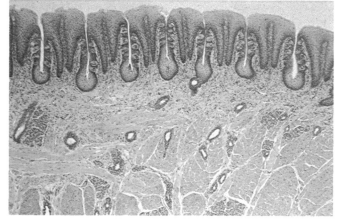

Fig. 253.

Fig. 253. The foliate papillae are only poorly developed in man, but well developed in a number of animal species such as the rabbit. At the posterolateral aspect of the tongue they form an oval area, the foliate region, consisting of slender mucosal folds oriented perpendicularly to the lingual border. The epithelium lining the trenches of these papillae is particularly rich in taste buds that at low magnifications appear as cone-shaped translucencies due to their poor stainability (for details cf, Figs. 447, 448). Iron-hematoxylin staining. Magnification 38×.

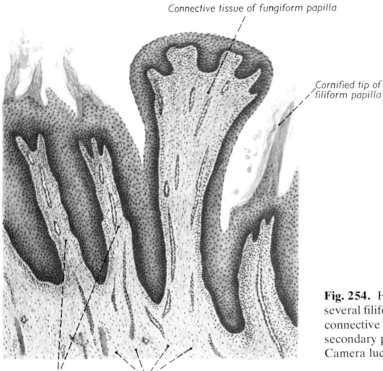

Connective tissue of fungiform papilla

Cornified tip of filiform papilla

Connective tissue core of filiform papillae

Lamina propria

Fig. 254. Higher magnification of the lingual mucosa with several filiform and a single fungiform papilla. Note that the connective tissue core (= primary papilla) subdivides into secondary papillae toward the epithelium (human tongue). Camera lucida drawing. H.E. staining. Magnification 60×.

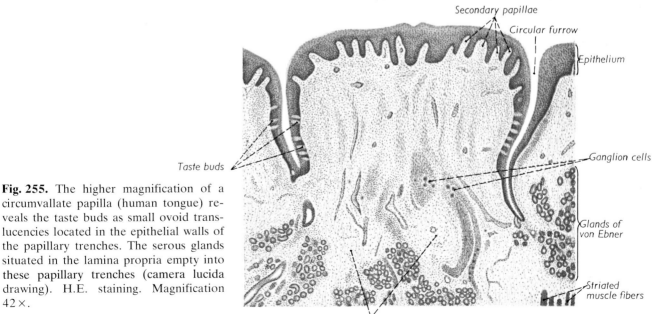

Secondary papillae

Circular furrow

Epithelium

Taste buds

Ganglion cells

Glands of von Ebner

Striated muscle fibers

Lamina propria

Fig. 255. The higher magnification of a circumvallate papilla (human tongue) reveals the taste buds as small ovoid translucencies located in the epithelial walls of the papillary trenches. The serous glands situated in the lamina propria empty into these papillary trenches (camera lucida drawing). H.E. staining. Magnification 42×.

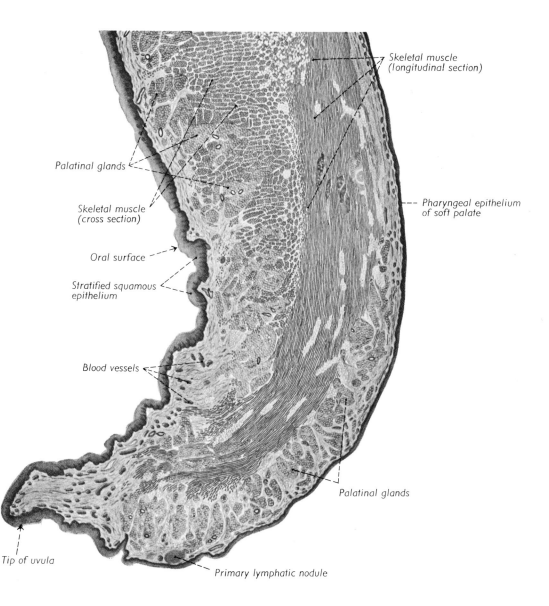

Skeletal muscle
(longitudinal section)

Palatinal glands

Skeletal muscle
(cross section)

Pharyngeal epithelium
of soft palate

Oral surface

Stratified squamous
epithelium

Blood vessels

Palatinal glands

Tip of uvula

Primary lymphatic nodule

Fig. 256. Longitudinal section through the soft palate and the uvula. As already described for the lips, the central tissue core of this specimen is also made mainly of striated muscle fibers. Unlike the lips, however, the cellular lining covering both its palatinal and pharyngeal surface is a noncornified stratified squamous epithelium that only differs in its height. Its continuation into the respiratory epithelium of the nasal cavity – contrary to a widespread assumption – is never found at the free margins but can be shifted so far onto the pharyngeal surface that it is not included in the section as in this specimen (camera lucida drawing). For detailed analysis and identification see Table 11. H.E. staining. Magnification 7.5 ×.

The three large salivary glands of the oral cavity, the parotid, the submandibular and the sublingual glands, differ in both the type of their secretory units and the organization and composition of their duct systems. The duct systems are particularly useful in discriminating these glands from other similar exocrine glands, such as the lacrimal gland or the pancreas (cf. Figs. 267, 268 and Table 12).

Fig. 257. Low-power view of the exclusively serous human parotid gland in which the large number of duct profiles – in this specimen predominantly striated (salivary) ducts – is particularly striking and can best be evaluated by means of the lowest microscopic objective. Furthermore, this large number of duct profiles is an essential criterion for distinguishing the parotid gland from the pancreas and the lacrimal gland. In the connective tissue between the serous alveoli, fat cells are often seen. These cells also occur in other salivary glands. A characteristic feature of the parotid gland, though not regularly found in every specimen, is the profiles of larger nerve bundles (ramifications of the facial nerve). Mallory-azan staining. Magnification 42 ×.

Fig. 258. Even at a low magnification the mixed human submandibular gland clearly exhibits both the different stainability of its secretory units and the less well-developed duct system when compared with the parotid gland. With the Mallory-azan stain the mucous tubules present a light bluish appearance, while with H. and E. they remain more or less unstained and hence appear "white." Mallory-azan staining. Magnific. 42 ×.

Fig. 259. The human sublingual gland is also a mixed gland but it is preponderantly mucous. Because of the great amount and the poor stainability of the mucous tubules, these can simulate for the beginner both a serous nature and an apparent homogeneity of the secretory units. Here also one of the characteristic features is the considerably reduced number of duct profiles when compared with the parotid gland. Mallory-azan staining. Magnification 42 ×.

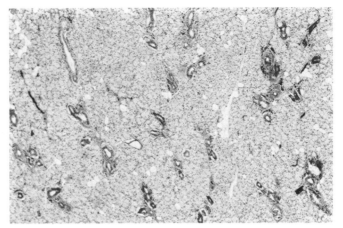

Fig. 257.

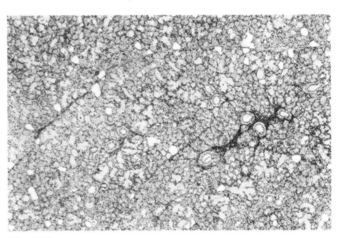

Fig. 258.

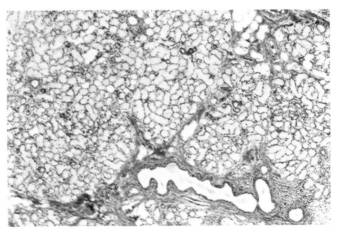

Fig. 259.

121

Salivary (striated) duct   Intercalated duct

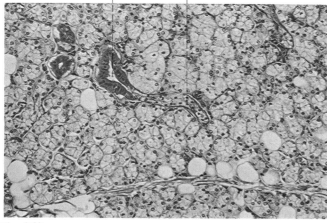

Fig. 260.

Fig. 260. The secretory units of the human parotid gland can only be identified at higher magnifications. They are of different sizes because they consist of a varying number of secretory cells. Their nuclei are never flattened, but regularly show a roundish outline and are often found at the cell base due to a massive accumulation of secretory products (cf. Figs. 81, 86, 263). In the center of the micrograph the continuation of a longitudinally sectioned intercalated duct into a cross-sectioned and deeper staining striated (salivary) duct can be seen. Mallory-azan staining. Magnification 150 ×.

Mucous tubules

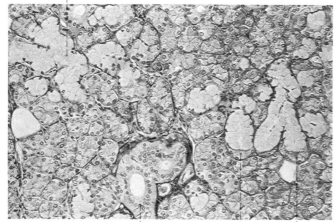

Fig. 261.

Fig. 261. In the submandibular gland the mucous tubules differ from the more or less berry-shaped serous secretory units (acini) not only by a different stainability but by their regularly flattened nuclei pressed against the cell base and by their tubular shape. In many instances the blind ends of the mucous tubules are capped by crescent-shaped groups of serous cells, the demilunes of von Ebner or crecents of Giannuzzi. Mallory-azan staining. Magnification 150×.

Salivary (striated) duct   Serous demilune

Serous alveolus

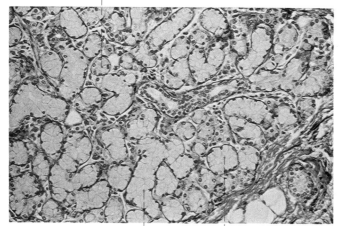

Fig. 262.   Mucous tubules

Fig. 262. The vast number of mucous tubules found in the human sublingual gland can easily simulate at first sight a homogeneity of the secretory units notwithstanding that both distinct serous demilunes and "free," i.e., serous alveoli not associated with the mucous portions, can be recognized. Mallory-azan staining. Magnification 150×.

122

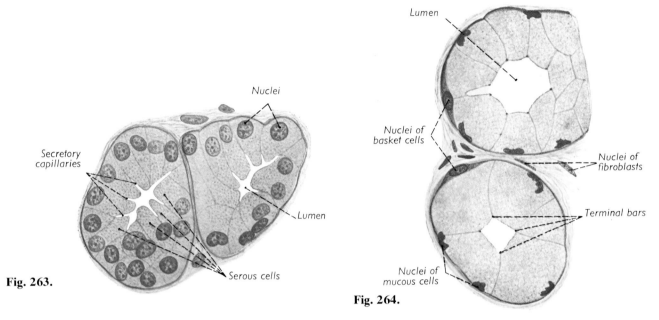

Fig. 263.

Nuclei

Secretory
capillaries

Serous cells

Lumen

Lumen

Nuclei of
basket cells

Nuclei of
fibroblasts

Terminal bars

Nuclei of
mucous cells

Fig. 264.

Fig. 263–265. Camera lucida drawings showing the cellular details of different secretory units from the human lingual gland (gland of Nuhn) at the same magnification (oil immersion, magnification 750 ×) and stained identically (H.E.). Only now the extremely narrow, often slit-shaped, lumina of the serous alveoli can be recognized (Fig. 263), whereas the secretory cells at lower magnification often seem to be apposed one to another without any interstices in between (cf. Fig. 260). Their nuclei always show a circular outline and are partly shifted toward the cell base. The lumina of the mucous tubules (Fig. 264) are usually much larger but are occasionally difficult to detect because the secretion contained within obscures the apical surfaces of the bordering (secretory) cells. The nuclei of the mucigenous cells are always flattened against the cell base and show an irregular outline.

In the mixed glands the serous cells very often engulf the blind ends of the mucous tubules in the shape of a crescent, the serous demilunes of von Ebner (Fig. 265). Their aqueous secretion dilutes and hence lowers the vicosity of the product expelled by the following mucous tubules and thereby enhances its flow velocity.

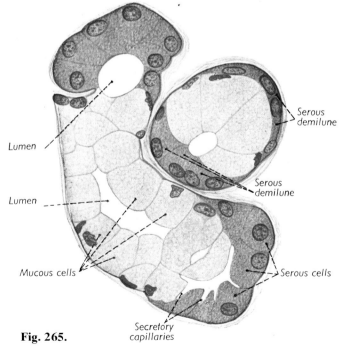

Serous
demilune

Lumen

Serous
demilune

Lumen

Mucous cells

Serous cells

Secretory
capillaries

Fig. 265.

Differential diagnosis of serous glands

Different segments of the duct system

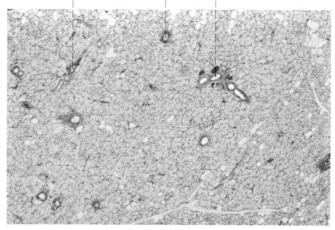

Fig. 266.

Excretory duct   *Islet*

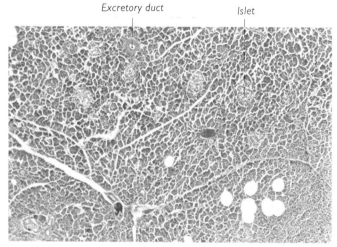

Fig. 267.

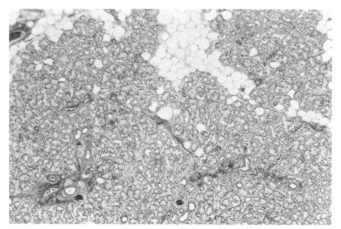

Fig. 268.

Figs. 266–268. The precise identification of the three large serous glands, i.e., the parotid gland (Fig. 266), the pancreas (Fig. 267) and the lacrimal gland (Fig. 268) can be best accomplished by the use of a low-power objective and not, as often believed, by means of delicate structural details, such as the centroacinar cells in the pancreatic secretory units.

The safest and nearly crucial criterion for identifying the parotid gland is the large number of duct profiles found in every such specimen. This gland can readily be distinguished from the other two glands by this means.

The lacrimal gland may best be identified by the rather large and hence prominent lumina of its secretory units which, in addition, are more loosely arranged than those of the pancreas and parotid gland.

Identification of the pancreas is based on (1) the islets of Langerhans (also best seen at a low magnification because then their lighter staining clearly distinguishes them from the exocrine portions), (2) the occurrence of only a few but interlobularly located excretory ducts (nothing more can be seen at such a low magnification) and (3) the very poorly developed interlobular connective tissue septa (compare, on the other hand, with the parotid and lacrimal gland). Even in cases where the islets are missing in a given section (the head and the uncinate process are nearly devoid of islets), the last two criteria are sufficient for an unequivocal identification of the pancreas. All figures: Mallory-azan staining. Magnification 42×.

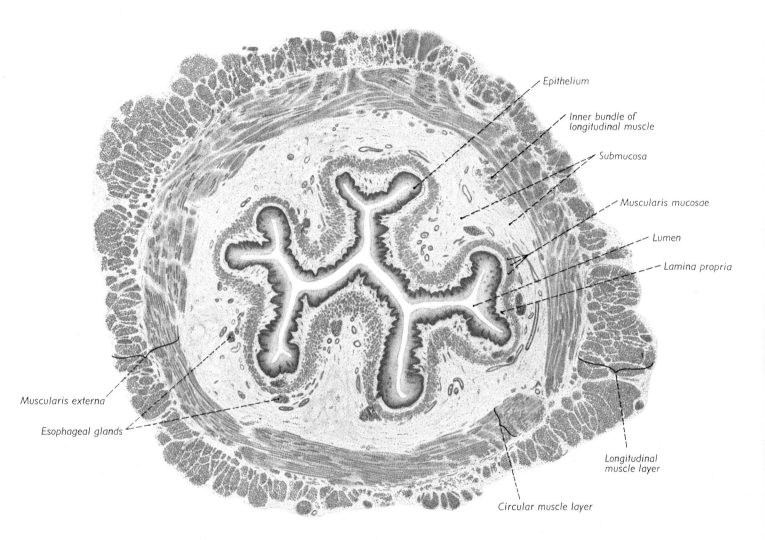

Epithelium

Inner bundle of
longitudinal muscle

Submucosa

Muscularis mucosae

Lumen

Lamina propria

Muscularis externa

Esophageal glands

Longitudinal
muscle layer

Circular muscle layer

Fig. 269. Camera lucida drawing of a complete cross section of a human esophagus to illustrate the typical structure of the wall that is maintained in a basically similar fashion throughout the remainder of the alimentary tract. It consists of (1) a mucosa comprising an epithelial lining together with a lamina propria made of loose and often reticular connective tissue, (2) a smooth muscle layer of various thickness, the muscularis mucosae, which is the most distinguishing feature of the "alimentary tract." This is followed by (3) the submucosa, which is surrounded by (4) the muscularis externa. The latter is regularly subdivided into an inner circular and an outer longitudinal layer, with the autonomic myenteric plexus in between. The appearance of these two muscular layers in a given specimen allows to decide whether one is confronted with a longitudinal or a cross section of the alimentary tract (in a *cross* section the inner circular layer is cut longitudinally).

By being endowed with a muscularis mucosae the esophagus not only clearly demonstrates that it belongs to the alimentary tract, but that it can also be definitely distinguished thereby from all the other regions showing an identical epithelium (stratified, noncornified and squamous) such as the oral cavity, the vagina, the cornea, the external urethral orifice and the uterine portio vaginalis. The esophagus can also be distinguished from the remainder of the alimentary tract by its epithelium, because, all the other parts possess a simple columnar epithelium. In case of doubt, the occurrence of small glands lying within the esophageal submucosa might ensure its differentiation from such structures as the vagina. But as these esophageal glands are rather widely spaced and few in number, they need not be found in every section, and their absence does not rule against the identification "esophagus" if all the other criteria are in favor of it. H.E. staining. Magnification 11 ×.

Gastric pits   Muscularis mucosae

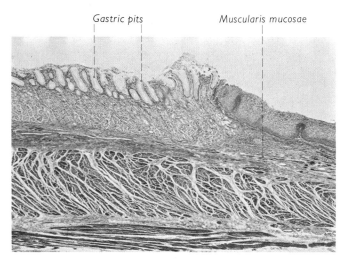

Fig. 270.

Fig. 270. Longitudinal section (because the outer muscular layer is cut longitudinally!) through the human cardia, the juncture of esophagus and stomach. Together with the characteristic and abrupt change from a stratified noncornified squamous epithelium to a simple columnar one the occurrence of epithelial crypts (= gastric pits) should be noted. They are continuous with the tubular gastric glands extending into the deeper mucosal layers. Mallory-azan staining. Magnification 19×.

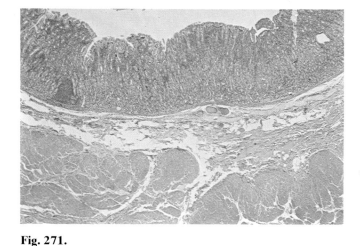

Fig. 271.

Fig. 271. Longitudinal section of a human gastric fundus with closely spaced secretory tubules in its mucosa. These empty into epithelial indentations (gastric pits) that here are relatively shallow when compared with the total mucosal height. The narrow muscularis mucosae is indistinguishable in this case because of the low magnification. A reliable distinction between the gastric fundus and the colon, with which it is often confused, is best accomplished by simply noting that goblet cells are never found in the gastric pits or glands, but do occur regularly in large numbers in the colic crypts (cf. Fig. 85). H.E. staining. Magnification 19×.

Gastric pits   Lymphatic nodule

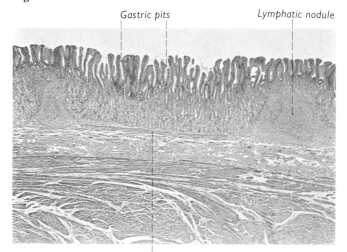

Fig. 272.   Muscularis mucosae

Fig. 272. When compared with the gastric fundus, the epithelial pits of the pyloric mucosa are much deeper, hence occupying a greater proportion of the entire mucosal thickness and therefore are more readily recognized. Lymphatic aggregations of various sizes can be seen among the secretory tubules of the pyloric glands as found in many other mucosal membranes. The largest of these (on the left side of the micrograph) shows a germinal center and therefore has to be classified as a lymphatic nodule (do not confuse with the aggregated lymphatic follicles of the ileum that are located within the submucosa). H. and chromotrop staining. Magnification 19×.

Besides possessing the usual four coats constituting the wall of the alimentary tract, the three successive parts of the small intestine, i.e., (1) duodenum, (2) jejunum and (3) ileum show a characteristic modeling of their mucosal surfaces, namely the simultaneous occurrence of both folds *and* villi. The "folds" (valves of Kerckring or plicae circulares) can easily be seen with the naked eye, and they involve not only the entire mucosa but also parts of the submucosa as well, which therefore constitute their central connective tissue core.

The "villi," however are much smaller, finger-shaped projections of the mucosal membrane alone, and can only be recognized by means of a magnifying glass or the lowest power objective. As the plicae circulares decrease considerably in number toward the ileum, specimens of this part of the small intestine often involve no folds at all, as seen Fig. 275. But this is no argument against identifying "ileum" if all the other criteria are in favor of it. In order to include in a single section as many of the circularly oriented folds as possible, specimens of the small intestine are usually cut longitudinally. If, however, transverse sections are used, folds could be totally missing, and it is therefore important to decide primarily in which plane the two layers of the muscularis externa are cut in a given specimen (for differential diagnosis see Table 13).

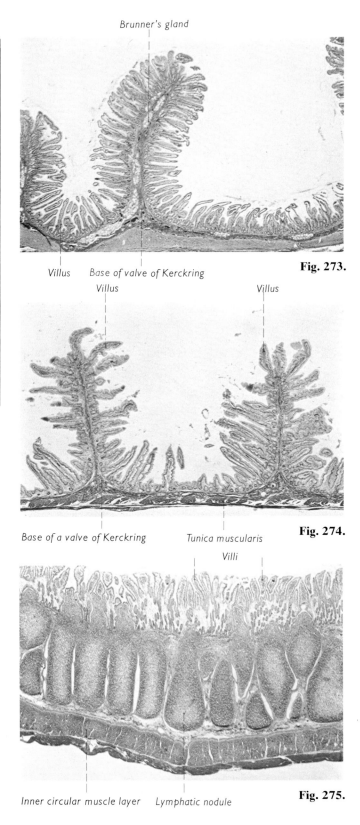

Brunner's gland

Villus   Base of valve of Kerckring   **Fig. 273.**

Villus   Villus   Villus

Base of a valve of Kerckring   Tunica muscularis   **Fig. 274.**

Villi

Inner circular muscle layer   Lymphatic nodule   **Fig. 275.**

Fig. 273. Longitudinal section of the human duodenum showing two folds that, like the remainder of the mucosal surface, are covered with closely spaced villi. Within the submucosa, including that of the valves of Kerckring, lightly staining areas corresponding to the glands of Brunner can be identified. These are the distinguishing features of "duodenum" and they clearly delineate this part of the small intestine from all the remainder. Mallory-azan staining. Magnification 12 ×.

Fig. 274. Longitudinal section through a human jejunum with two closely spaced folds, whose submucosa as well as that of the remaining intestinal wall is completely devoid of glands. Due to a variant degree of shrinkage several villi show translucent clefts of various widths, located between their epithelium and the connective tissue core. Mallory-azan staining. Magnification 21.5 ×.

Fig. 275. Though cut longitudinally, none of the widely spaced folds has been included in this specimen of a human ileum. Its closely spaced villi only occasionally exhibit some larger artificial clefts. The most characteristic and hence essential feature for identification is the aggregation of lymphatic nodules (Peyer's patches) located within the submucosa. Mallory-azan staining. Magnification 14.5 ×.

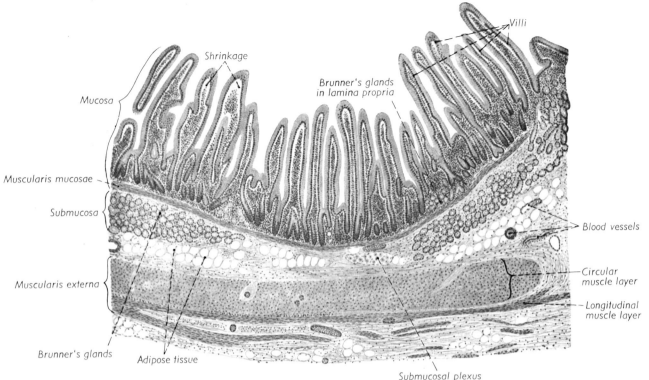

Fig. 276. At a higher magnification of the duodenal mucosa (man) it becomes evident that the epithelium (simple, columnar) not only covers the villous projections but also lines tubular indentations (= crypts of Lieberkühn) that with their blind ends reach the muscularis mucosae. These crypts not only occur in the duodenum, but in the remaining parts of the small intestine as well, and hence this portion of the digestive tube is characterized by (1) folds, (2) villi and (3) crypts. H.E. staining. Magnification 50×.

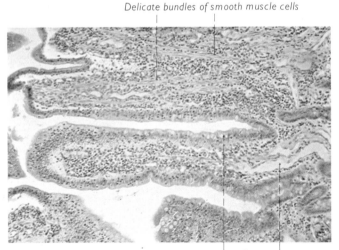

Fig. 277. A higher magnification of the jejunal villi from at cat clearly shows the goblet cells as lighter staining ellipsoids scattered throughout the entire intestinal epithelium. In addition delicate strands of smooth muscle become apparent that project from the muscularis mucosae into the connective tissue core of the villi. These elements are responsible for the retraction of the villi during their motile activity known as the "villous pump." The intestinal lumen is at the left edge of the micrograph. H.E. staining. Magnification 96×.

Fig. 278. As biopsies taken for diagnostic purposes from different parts of the digestive canal are also studied with the electron microscope, a semi-thin section (0.5–1 μm) as it is routinely prepared in the course of this technique is shown in this micrograph from a rat duodenum. The striated border appears with the stain applied as a dark blue line, while the goblet cells stand out clearly as deeply colored round or ovoid corpuscles. The intestinal lumen lies outside of the left edge of the micrograph. Methylene blue-azur II staining. Magnification 150×.

Parietal cell

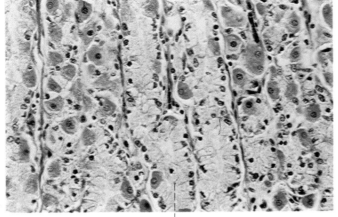

Fig. 279.   Chief cells

Smooth muscle cells in lamina propria of a villus

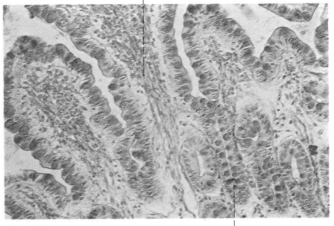

Fig. 280.   Crypt of Lieberkühn

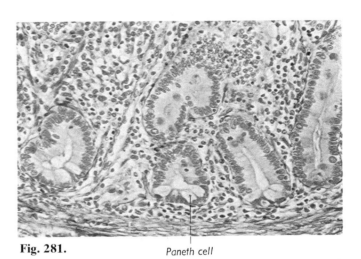

Fig. 281.   Paneth cell

Reticular connective tissue of lamina propria

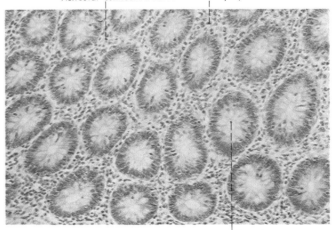

Fig. 282.   Lumen of crypt

Fig. 279. Unlike the pyloric glands the secretory tubules of the human gastric fundus are equipped with the acidophilic parietal cells. In this specimen they appear as brownish-red elements attached to the outer surfaces of the glandular tubules secreting the ionized hydrogen necessary for the synthesis of hydrochloric acid. As the parietal cells are often difficult to see due to a faded and/or a rather nonspecific staining, e.g., H. and E., the identification of "gastric fundus" has to be based on the structure of the entire mucosa (cf. Fig. 271). The apparent absence of the parietal cells is no argument against identifying the specimen as "gastric fundus" if all the other morphological features are strongly in favor of it. Iron-hematoxylin and thiazine red staining. Magnification 240×.

Fig. 280. Villi and crypts of the human ileum in whose epithelial lining numerous goblet cells (stained blue) are interspersed. Note within the villous stroma the elongated slender smooth muscle cells (particularly clear in the center of the micrograph) that are loosely aggregated and deviate from each other like a fountain when approaching the top of the villus. Mallory-azan staining. Magnification 150×.

Fig. 281. Obliquely, transversely and longitudinally sectioned crypts lying within the lamina propria of the human duodenal mucosa. In their depths can be seen the cells of Paneth which are assumed to be secretory elements, but their specific product is still not well established. Mallory-azan staining. Magnification 240×.

Fig. 282. A nearly ideal cross section through the crypts of the colic mucosa that never shows any epithelial projections (villi) but only these regularly spaced invaginations. In contrast to cross-sectioned villi, here the epithelium surrounds a central opening, the lumen of the crypt, whereas it would encircle a connective tissue core in a transverse section of a villus. The goblet cells appear as oval-shaped translucencies within the epithelial lining. H.E. staining. Magnification 150×.

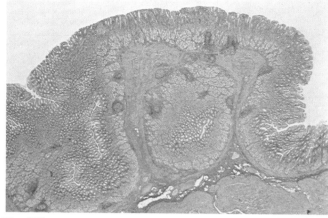

Fig. 283.

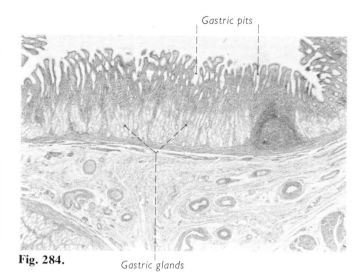

Gastric pits

Fig. 284.

Gastric glands

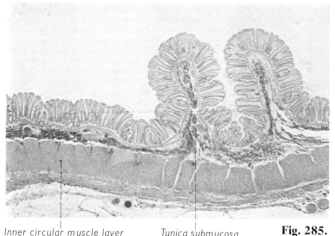

Inner circular muscle layer   Tunica submucosa   **Fig. 285.**

Fig. 283–286. A comparison of various portions of the alimentary tract that are often confused with each other and/or misinterpreted. Note that the stomach also might show folds (Fig. 283), but these are much coarser than those found in the small intestine. The pyloric portion (Fig. 284) can be distinguished from the fundus (Fig. 283) by (1) its deeper gastric pits occupying a greater proportion (one half) of the mucosal height and (2) its more loosely packed tubular glands. Finally the entire stomach can be distinguished from the colon (Fig. 285) with which it is often confused by its considerably greater total thickness, particularly of its muscularis externa. The large intestine (Fig. 285) occasionally might have folds, but its mucosa consists exclusively of regularly arranged crypts in whose epithelium numerous goblet cells are interspersed (cf. Fig. 85). Thus the colon possesses neither the mucosal villi characteristic of the small intestine nor the elongated branched tubular glands of the stomach.

The gall bladder (Fig. 286) often remains unrecognized because it is not taken into consideration at all when one is concerned with the identification of the various parts of the alimentary tract to which it belongs only in a wider sense. The gall bladder is characterized (1) by the absence of a muscularis mucosae, (2) by a muscularis not subdivided into two distinct layers and (3) by numerous irregular and narrow mucosal folds. As the last are interconnected with each other, leaving irregular polygonal depressions in between, a section through the gall bladder must give the impression of "anastomosing" folds enclosing epithelial cavities of various sizes. H.E. staining (Fig. 283, 284 and 286), Mallory-azan staining (Fig. 285). Magnifications 9–, 11–, 13.5 and 48× respectively.

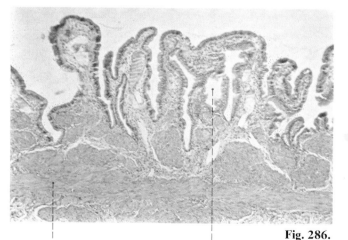

Tunica muscularis   Epithelial diverticulum   **Fig. 286.**
                           or crypt

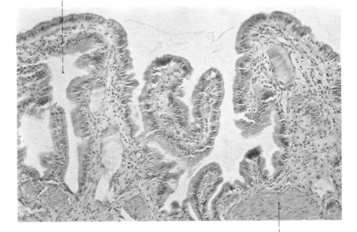

Epithelial diverticulum in a mucosal fold

Fig. 287. At a higher magnification it can be seen that the simple columnar epithelium consists of particularly tall cells and is completely devoid of goblet cells. The muscularis is made of interlacing bundles of smooth muscle and separated from the epithelial covering by a poorly defined lamina propria (human gall bladder). H.E. staining. Magnification 96×.

Innermost layer of muscular tunic

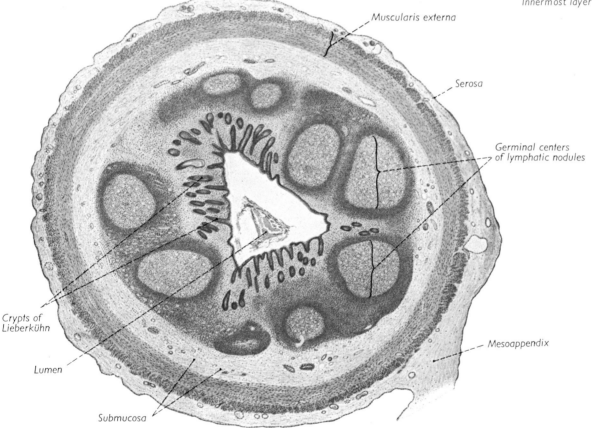

Muscularis externa

Serosa

Germinal centers of lymphatic nodules

Crypts of Lieberkühn

Lumen

Submucosa

Mesoappendix

Fig. 288. Complete transverse section through the vermiform appendix whose mucosa closely resembles that of the colon except that its crypts are not so regularly spaced and may be partially missing. Particularly striking are the numerous lymphatic nodules scattered throughout the entire lamina propria and reaching into the submucosa. Thereby they not only displace the crypts to a various extent, but they also infiltrate and disrupt the thin muscularis mucosae and thereby make this layer difficult to identify in the vermiform appendix (camera lucida drawing). H.E. staining. Magnification 22×.

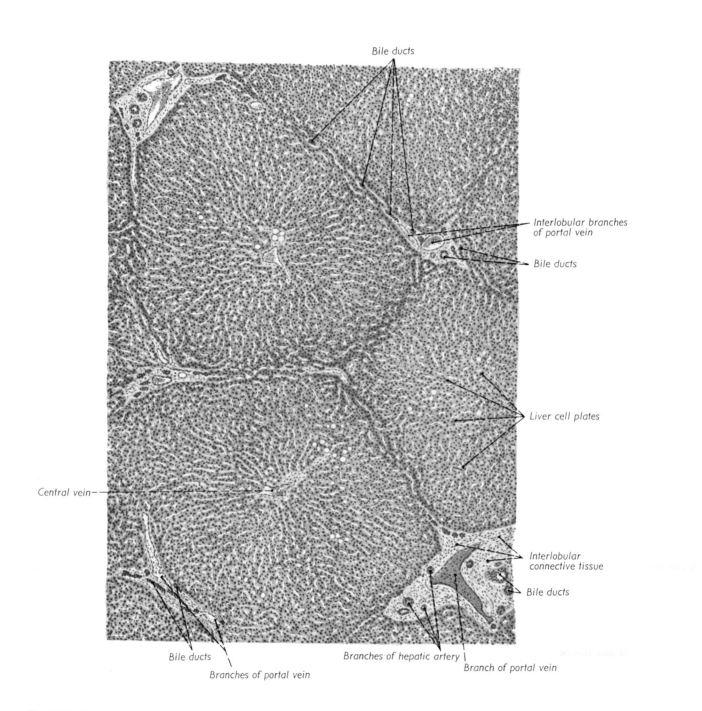

Bile ducts

Interlobular branches
of portal vein

Bile ducts

Liver cell plates

Central vein

Interlobular
connective tissue

Bile ducts

Bile ducts

Branches of portal vein

Branches of hepatic artery

Branch of portal vein

Fig. 289. The organization of the liver into innumerable "hepatic lobules" has been artificially emphasized in this camera lucida drawing in order to outline these structural units more clearly than usually seen in a human liver. In a section these lobules appear as small, more or less roundish or irregular polygonal units consisting of (1) hepatic cells arranged in plates or cords radiating around a central blood vessel (central vein) and (2) the interconnecting liver sinusoids between. The interlobular connective tissue is apparent mainly at points where three or more lobules meet to form the portal area or canal. These regularly contain the interlobular bile duct together with branches of the portal vein and the hepatic artery. The duct and two types of vessels are known as the "portal triad," which lies in and constitutes the main constituents of the portal canal. H.E. staining. Magnification 70 ×.

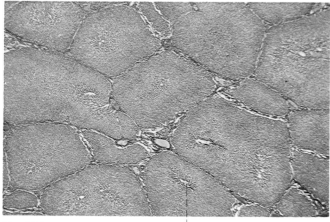

Fig. 290.   *Central vein*

Central vein

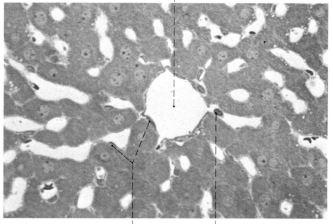

Fig. 291.   *Endothelial nuclei*   *Nucleus of a Kupffer cell*

Central vein

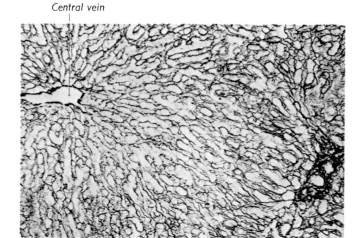

Fig. 292.

Fig. 290. Low-power view of a porcine liver showing a particularly clear delineation of the hepatic lobules due to their complete investment by connective tissue septa, a chracteristic of this species. Because the porcine liver shows the hepatic lobules so clearly, it is often chosen as the specimen to be studied first by the beginning student, Mallory-azan staining. Magnification 19×.

Fig. 291. Center of a rat hepatic lobule in a semi-thin section as routinely prepared from biopsies in modern liver diagnostics. Due to the thinness of such sections (0.5–1μm) and to a better fixation (in this experimental animal by vascular perfusion), these specimens reveal many more structural details than those obtained by the older techniques. Note distinct endothelial nuclei in the central vein and in the sinusoids. Methylene blue-azur II staining. Magnification 380×.

Fig. 292. Detail of a human hepatic lobule showing the delicate network of reticular fibers enmeshing the hepatic cells, whose radiating arrangement around the central vein is clearly outlined in such preparations. Silver impregnation. Magnification 95×.

Fig. 293. Portal canal (area) from a human liver with the "portal triad" whose individual components, i.e., an interlobular bile duct together with a branch of the portal vein and of the hepatic artery, can be easily distinguished by the different structure of their walls. H.E. staining. Magnification 150×.

Hepatic artery

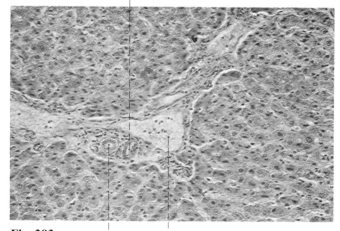

Fig. 293.   *Bile duct*   *Interlobular branch of portal vein*

133

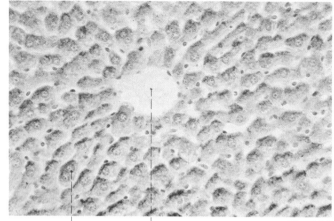

Binucleated hepatic cell   Central vein   **Fig. 294.**

Vacuoles within hepatic cells   *Kupffer cell*

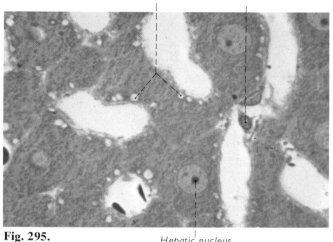

Fig. 295.   *Hepatic nucleus*

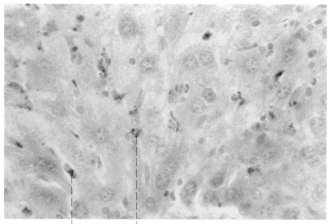

Kupffer cells   **Fig. 296.**

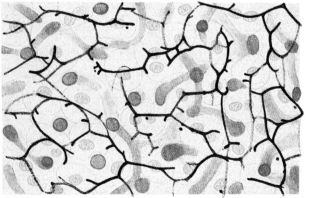

Fig. 297.

Fig. 294. Hepatic cell cords showing the intracellular deposits of glycogen in form of granules of various sizes (cf. Fig. 40). PAS and hemalum staining. Magnification 240×.

Fig. 295. Semi-thin section of liver parenchyma showing several sinusoids with the adjoining hepatocytes that display numerous vacuoles along their surfaces towards the space of Disse. Note the delicate endothelium and the nucleus of a Kupffer cell within the longitudinally sectioned sinusoid (right side of the micrograph). Compare also with Figs. 299 and 300. Methylene blue-azur II staining. Magnification 960×.

Fig. 296. The stellate cells of von Kupffer are located within the liver sinusoids and they belong to the reticuloendothelial system. They can only be seen with the light microscope by utilizing their highly phagocytic activity and thereby "marking" these elements with ingested foreign materials, e.g., intravitally injected trypan blue. Note the binucleate hepatic cell. Staining: Trypan blue intravitally and nuclear fast red. Magnification 600×.

Fig. 297. Illustration of the tridimensional network of the bile "capillaries" with a silver impregnation technique (camera lucida drawing). These minute tubules (= canaliculi) have no "wall" of their own but are formed as recesses between adjacent hepatic cells whose plasmalemmata must, therefore, function also as the linings of these bile canaliculi (cf. Fig. 301). Silver impregnation and alum-carmine. Magnification 380×.

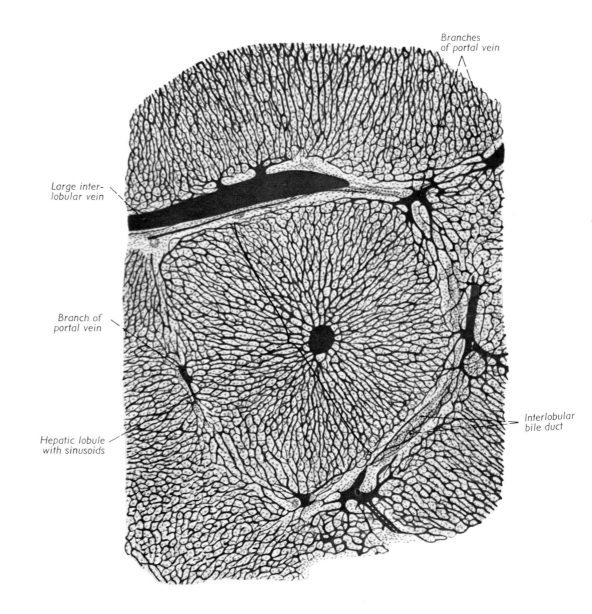

Branches
of portal vein

Large inter-
lobular vein

Branch of
portal vein

Interlobular
bile duct

Hepatic lobule
with sinusoids

Fig. 298. Injection specimen of a rabbit's liver (camera lucida drawing). To demonstrate the entire vascular bed the organ has been perfused via the portal vein with a colored (Berlin blue) gelatin solution. For a similar purpose double-injected specimens are often used in which differently stained solutions are simultaneously injected via the hepatic *and* the portal vein, thereby trying to delineate the two venous systems that meet in the hepatic lobule by a different coloring (the central veins represent the beginnings of the hepatic venous system). Borax-carmine staining. Magnification 54×.

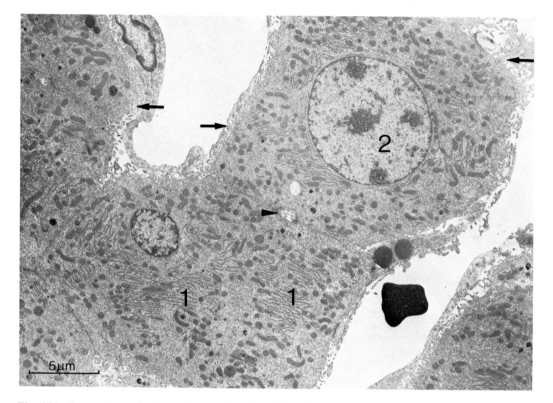

Fig. 299. Low-power electron micrograph of a rat liver illustrating two sinusoids together with the adjoining hepatocytes. These are characterized by an abundance of mitochondria and regular stacks of rough endoplasmic reticulum (1). Their surface toward the space of Disse is covered by closely spaced yet irregular microvilli (→) and a few of these project into the bile canaliculi (▶). 2 = Nucleus of a liver cell. Magnification 3,500×.

▶

Fig. 300. Transverse section through a sinusoid with a lymphocyte in its lumen (rat liver), revealing the extremely attenuated endothelium. This appears over large areas as a row of small insulated cytoplasmic profiles, and it forms only occasionally a continuous cellular layer over shorter distances (1). Note the large number of microvilli projecting into the space of Disse. Magnification 14,000×.

Fig. 301. Three hepatocytes bordering upon a bile canaliculus (*) that has no wall of its own and shows a few irregular microvilli in its lumen. One of the adjoining liver cells clearly exhibits parts of its nucleus (1), a regularly arranged rough ER (2), numerous mitochondria and glycogen particles (3). Magnification 14,000×.

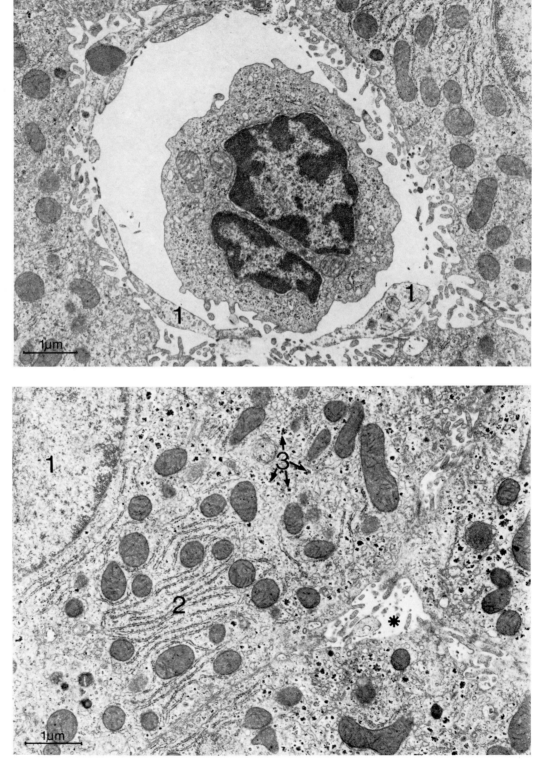

Fig. 300.

Fig. 301.

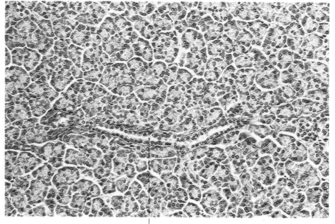

Small interlobular duct

Fig. 302. Exocrine portion of a human pancreas showing parts of a narrow interlobular duct. Though the islets of Langerhans are missing in this particular area, an exact identification of the pancreas and its distinction from the other serous glands can easily be achieved (for this cf. Fig. 266, 267 and Table 12). Mallory-azan staining. Magnification 150×.

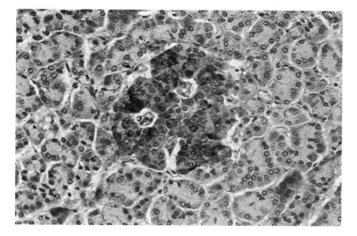

Fig. 303. Pancreatic islet of a dog with the ß-granules selectively stained by chromium hematoxylin. With this technique A- and B-cells can be distinguished from each other and their numerical proportions can be established. Chromium hematoxylin-phloxine staining. Magnification 280×.

Centro-acinar cells

Intercalated duct
(longitudinal section)

Capillary

Islet of
Langerhans

Interstitial
connective tissue

Capillary
in exocrine
pancreas

Exocrine
secretory
units

Interlobular
duct

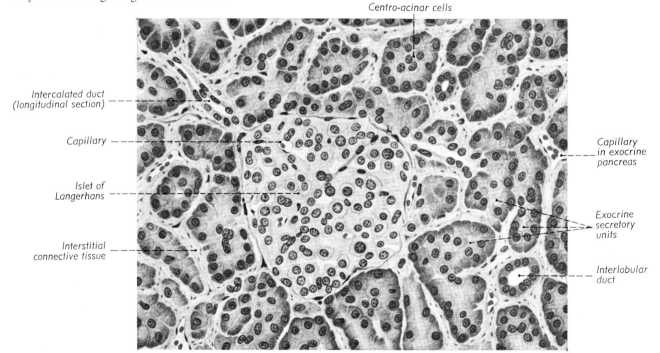

Fig. 304. A higher magnification reveals the basophilic substance found in the basal zones of the secretory cells that represents the light microscopical equivalent of the ergastoplasm (camera lucida drawing). The "centro-acinar" cells occur because the extremely narrow intercalated ducts are deeply invaginated into the secretory units, and hence their lining epithelial cells apparently lie in the center of the serous acini (= centro-acinar). But this morphological feature is only of limited value as a criterion for identificaton because the inexperienced microscopist often will be unsure about its definite identification. The islets of Langerhans can best be found with a low magnification because then their lesser stainability outlines them as roundish light areas within the exocrine glandular tissue. For the various cell types of the islets see Fig. 303. H.E. staining. Magnification 500×.

138

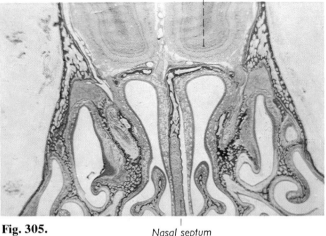

Fig. 305.

Olfactory bulb

Nasal septum

Fig. 305. Frontal section through the upper parts of a feline nasal cavity which, as in all the other species with a highly developed olfactory sense, shows a much more elaborate system of conchae than that found in man. Note cross-sectioned olfactory bulbs at the top of the micrograph. Mallory-azan staining. Magnification 10×.

Epidermis   Sebaceous gland

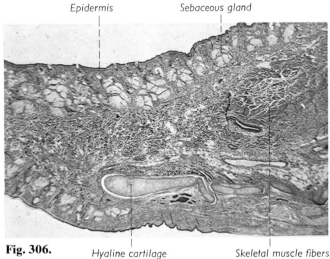

Fig. 306.   Hyaline cartilage   Skeletal muscle fibers

Fig. 306. Sections of the nasal ala are characterized by an outer covering of skin that contains sebaceous glands but no hairs, and an inner surface lined similarly but showing thick hairs (vibrissae). Other areas of the inner surface further inside and away from the hairs may be covered by a respiratory epithelium. The central tissue core is made of both hyaline cartilage and skeletal muscle fibers in varying proportions. For detailed identifying characteristics, see Table 11. Mallory-azan staining. Magnification 10×.

Ciliated epithelium

Lamina propria

Nasal glands

Veins

Bone

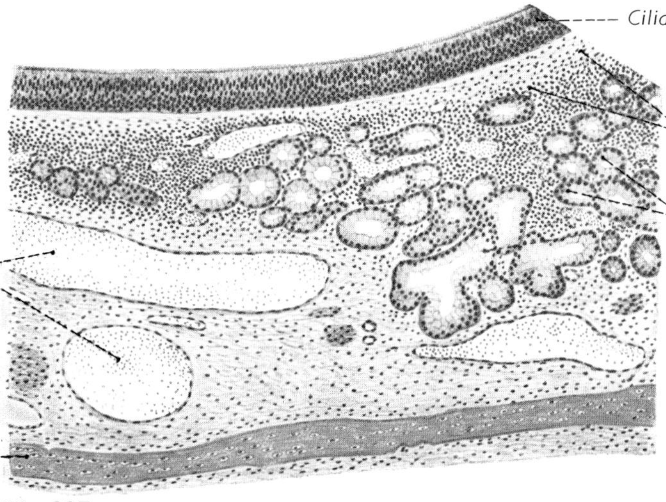

Fig. 307.

Fig. 307. The respiratory region of the nasal mucosa is covered by its characteristic pseudostratified ciliated columnar epithelium, and its lamina propria contains numerous tubulo-acinar (serous and mucous) glands together with many large veins (camera lucida drawing). H.E. staining. Magnification 110×.

139

Respiratory system – The epiglottis

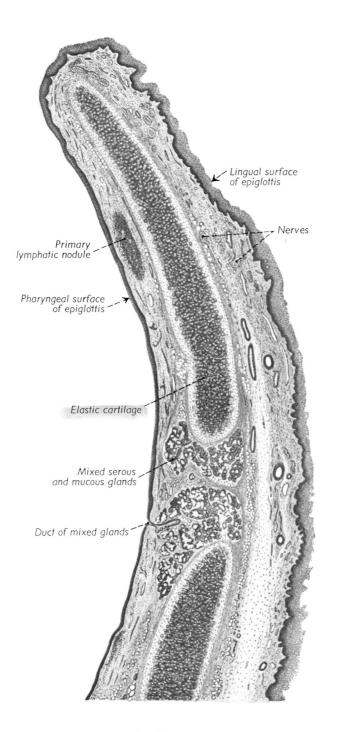

Lingual surface of epiglottis

Nerves

Primary lymphatic nodule

Pharyngeal surface of epiglottis

Elastic cartilage

Mixed serous and mucous glands

Duct of mixed glands

Fig. 308. Longitudinal section through a human epiglottis whose surfaces are covered by a stratified, noncornified and squamous epithelium of different heights (camera lucida drawing). The juncture with the respiratory epithelium is never found at the apex of the epiglottis but is often shifted so deep down that it is not included in the specimen as in this figure. The central tissue core is preponderantly represented by an elastic cartilage. For further identifying characteristics, see Table 11. H.E. staining. Magnification 16.5×.

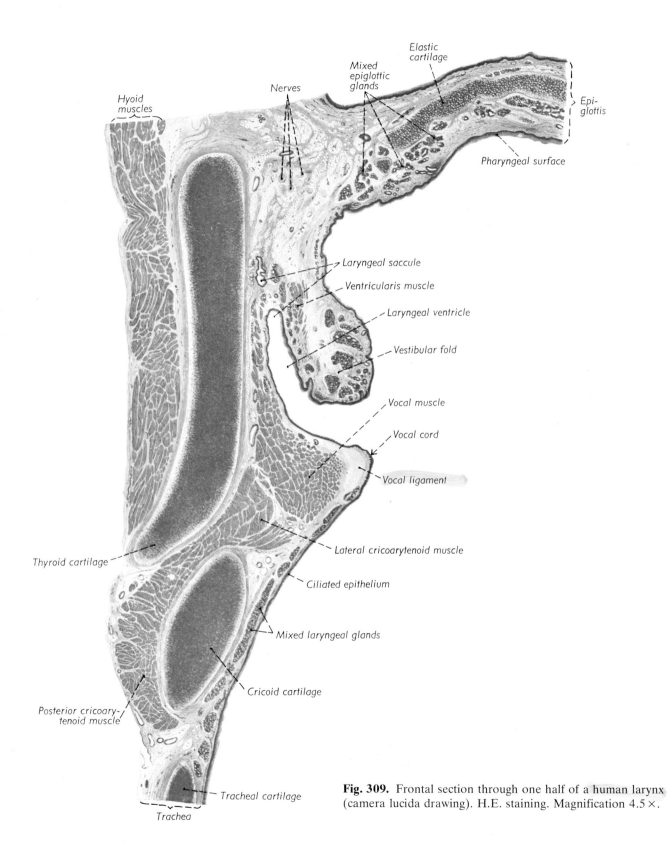

Hyoid muscles

Nerves

Mixed epiglottic glands

Elastic cartilage

Epiglottis

Pharyngeal surface

Laryngeal saccule

Ventricularis muscle

Laryngeal ventricle

Vestibular fold

Vocal muscle

Vocal cord

Vocal ligament

Lateral cricoarytenoid muscle

Thyroid cartilage

Ciliated epithelium

Mixed laryngeal glands

Cricoid cartilage

Posterior cricoary-tenoid muscle

Tracheal cartilage

Trachea

Fig. 309. Frontal section through one half of a human larynx (camera lucida drawing). H.E. staining. Magnification 4.5×.

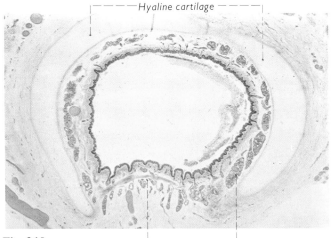

Fig. 310.

Hyaline cartilage

Trachealis muscle   Glands

Fig. 310. Cross section through a human fetal trachea that already shows all the structural features characteristic for the fully developed state. Its C-shaped hyaline cartillage is not cut a uniform width because this section is slightly oblique. Note the large number of small glands within the lamina propria and the trachealis muscle traversing the paries membranaceus. H.E. staining. Magnification 14×.

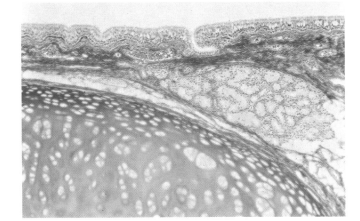

Fig. 311. Longitudinal section through a trachea of an adult showing its mucosa to consist of a ciliated pseudostratified columnar (respiratory) epithelium and an underlying fibrous lamina propria. The sero-mucous glands are preferably located within the interstices between the hyaline cartilages. Mallory-azan staining. Magnification 62.5×.

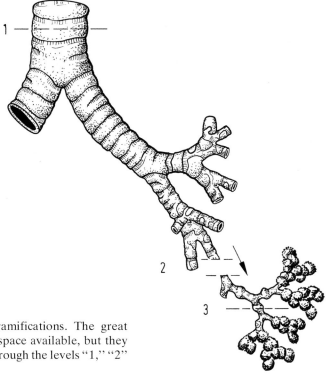

Fig. 312. Extremely simplified diagram of the trachea and its ramifications. The great number of intrapulmonary bronchi have been curtailed due to the space available, but they would have been located between the levels "2" and "3." Sections through the levels "1," "2" and "3" correspond approximately to Figs. 310, 313a and 314.

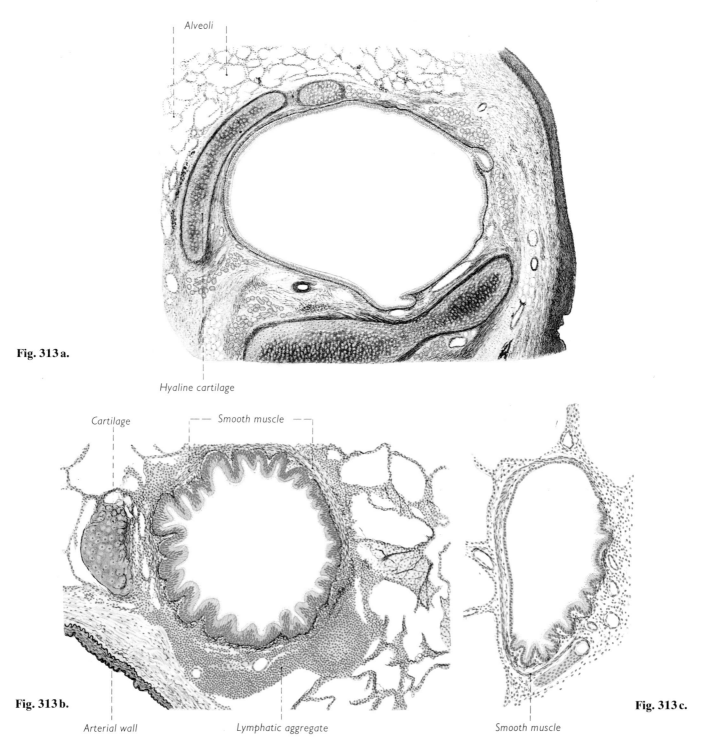

Alveoli

Fig. 313 a.

Hyaline cartilage

Cartilage — — *Smooth muscle* — —

Fig. 313 b.

Fig. 313 c.

Arterial wall   *Lymphatic aggregate*   *Smooth muscle*

Fig. 313. Cross sections through different parts of the intrapulmonary bronchial tree. Elastica and nuclear fast red staining. Magnifications 30× and 60× (313b, c).
a) Small bronchus with considerable amounts of hyaline cartilage in its wall.
b) Terminal branch of a bronchus showing sparse cartilaginous tissue together with a lymphatic infiltration.
c) Bronchiole displaying no cartilage but smooth muscle oriented obliquely and circularly.

Terminal bronchiole

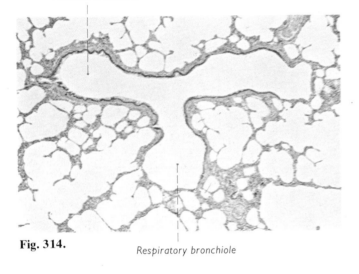

Fig. 314.

Respiratory bronchiole

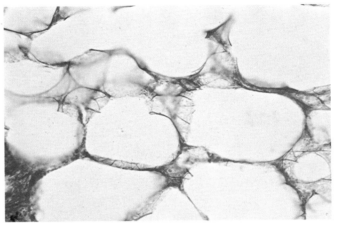

Fig. 317.

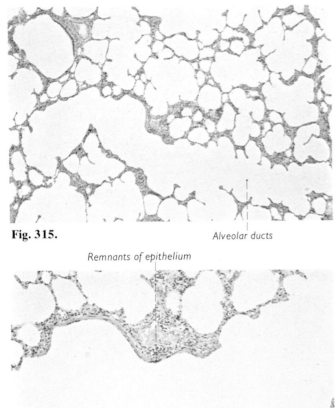

Fig. 315.

Alveolar ducts

Remnants of epithelium

Fig. 316.

Fig. 314. Site of the ramification of a terminal bronchiole that continues into two respiratory bronchioles. Note the abrupt ending of the epithelium at the beginnings of the respiratory bronchioles that – due to the plane of sections – appear as blind pouches (human lung). H.E. staining. Magnification 38 ×.

Fig. 315. Longitudinal sections through the continuation of a respiratory bronchiole into an alveolar duct (right side of the micrograph). A cross-sectioned terminal bronchiole can be seen in the left upper corner (human lung). H.E. staining. Magnification 38 ×.

Fig. 316. Close-up of the site of continuation from the preceding micrograph. The epithelium no longer forms a continuous layer but appears as a short row of nuclei over shorter distances with small bundles of smooth muscle beyond. H.E. staining. Magnification 96 ×.

Fig. 317. Alveoli specifically stained for elastic fibers. These form a complex tridimensional network of which a routine section can give but a rough impression. This disadvantage can be improved by studying thick sections in which this network can be followed through by making extensive use of the fine adjustment of the microscope (camine lung). Orcein staining. Magnification 96 ×.

Bronchus   Hyaline cartilage

Mesenchymal interstitial connective tissue

Fig. 318.

Fig. 319.

Fig. 318. As an epithelial derivative the fetal lung consists at certain developmental stages of tubular cavities lined by an epithelium that show numerous dichtotomous ramifications fitted with acinous endings. Therefore the fetal lung closely resembles certain glands with which it can easily be confused (for differentiation cf. Fig. 375 and Table 14). Mallory-azan staining. Magnification 38×.

Fig. 319. At a higher magnification the homogeneity and thereby the low degree of differentiation of the cellular linings of all these cavities is evident. In addition the large number of cells lying in the interstitial connective tissue points to its mesenchymal character. Mallory-azan staining. Magnification 150×.

Fig. 320. The relatively narrow lung capillaries can only be identified with certainty in routine preparations when the specimens contain red blood cells. Only when they are artificially filled with a colored gelatin solution does one get information about the density of the capillaries and their tridimensional basket-like arrangement around the alveoli (feline lung). Injection with Berlin blue gelatin via the pulmonary artery, no counterstaining. Magnification 95×.

Fig. 321. Semi-thin approx. 1μ m thick, section of a feline lung that clearly shows the high capillary density even without any artificial filling of the microvasculature. In addition it reveals the delicacy of the air-blood barrier that is illustrated to a better advantage in the following electron micrographs. Methylene blue-azur II staining. Magnification 960×.

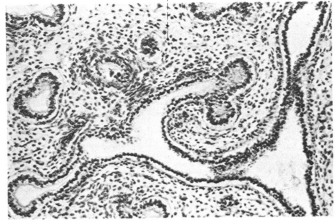

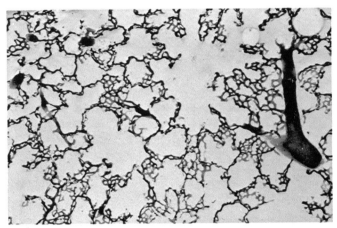

Fig. 320.

Capillary

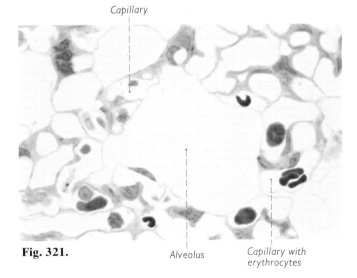

Fig. 321.   Alveolus   Capillary with erythrocytes

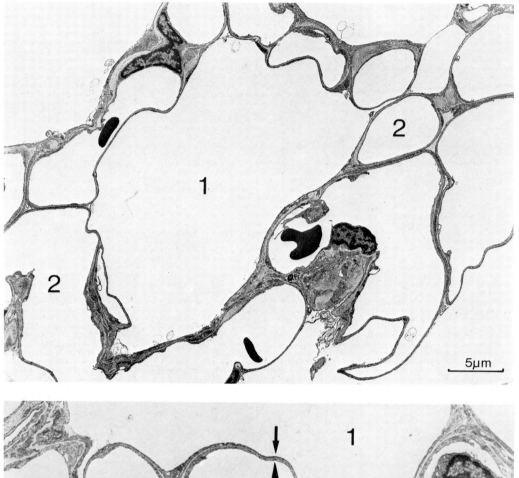

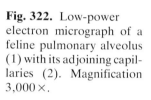

Fig. 322. Low-power electron micrograph of a feline pulmonary alveolus (1) with its adjoining capillaries (2). Magnification 3,000×.

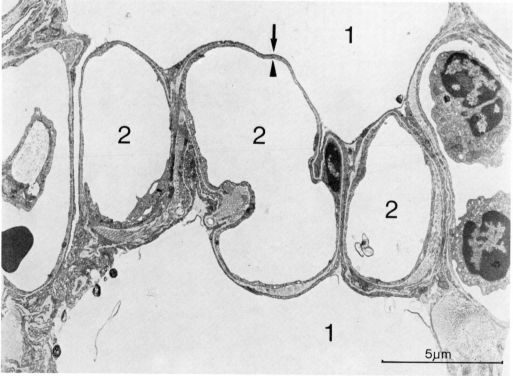

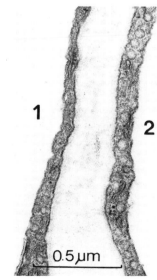

Fig. 323a.

Fig. 323b.

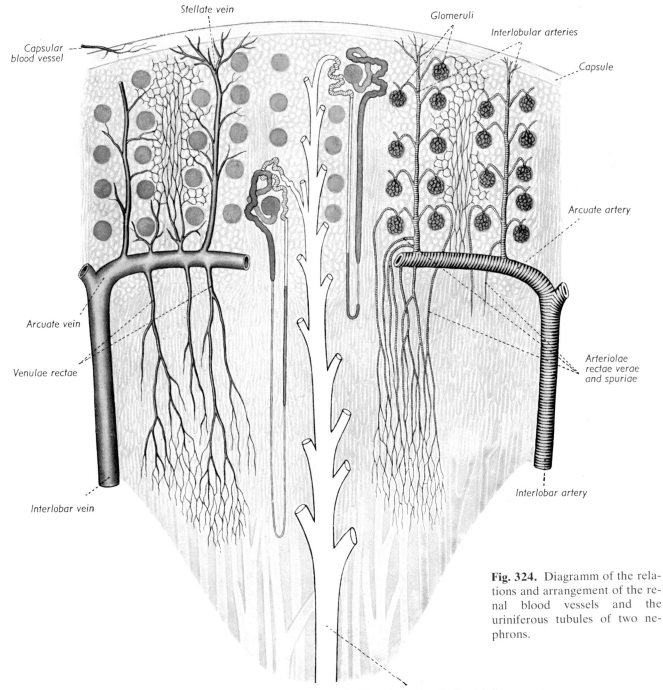

Capsular blood vessel

Stellate vein

Glomeruli

Interlobular arteries

Capsule

Arcuate artery

Arcuate vein

Venulae rectae

Arteriolae rectae verae and spuriae

Interlobar vein

Interlobar artery

Fig. 324. Diagramm of the relations and arrangement of the renal blood vessels and the uriniferous tubules of two nephrons.

Papillary duct with collecting tubules

Fig. 323. a) Profiles of three capillaries (2) located within an interalveolar septum of a feline lung. Note the prominent attenuation of the air-blood barrier that at these sites consists of only two delicate cellular layers, the alveolar epithelium ($\longrightarrow$) and the capillary endothelium ($\blacktriangleright$) together with a narrow intervening interstitial space. 1 = Alveolar lumen. Magnification 6,500×. b) A higher magnification better illustrates the individual components of the air-blood barrier that in this case contains a relatively broad interstitial space measuring $0.25\,\mu$m (= 250 nm). 1 = Alveolar lumen; 2 = Capillary lumen. Magnification 47,000×.

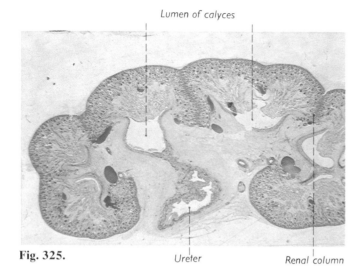

Fig. 325.

Lumen of calyces

Ureter

Renal column

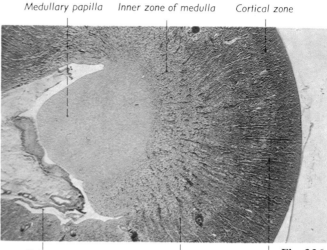

Medullary papilla   Inner zone of medulla   Cortical zone

Calyx   Outer zone of medulla   Medullary ray   Fig. 326.

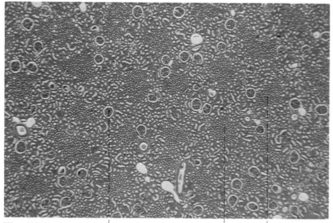

Fig. 327.   Glomerulus surrounded by cortical labyrinth   Medullary rays (cross section)

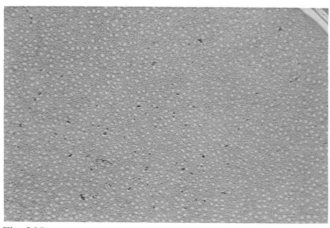

Fig. 328.

Fig. 325. Longitudinal section through a human fetal kidney (18 cm C.R. length) dividing the organ into two identical halves that clearly show a subdivision of the organ into lobes (= renculi) to each of which belongs a medullary pyramid and hence one minor calyx. The entire kidney parenchyma partially surrounds a cavity – the renal sinus – that in this specimen is mainly filled with connective tissue but additionally contains the renal pelvis and its major and minor ramifications (calyces) and the larger branches of the renal arteries, veins, nerves and lymphatics. Mallory-azan staining. Magnification 10×.

Fig. 326. Low-power view of a cross-sectioned rabbit's kidney with its parenchyma clearly showing an outer deeper staining cortex, whose outermost layer is colored even more intensely, and an inner medulla that is subdivided into different zones. Its innermost portion, the papilla, stains but poorly and is followed by an inner and outer medullary zone. As the latter continues into the cortex in the form of radiating strands, the medullary rays, no definite borders between these different parts of the renal parenchyma can be found. Mallory-azan staining. Magnification 10×.

Fig. 327. In a tangential section through the renal cortex (man) its subdivision by the medullary rays is clearly seen. These mainly consist of the straight portions of the proximal and distal tubules and are surrounded by areas known as the cortical labyrinth containing the convoluted tubules and renal corpuscles. H.E. staining. Magnification 24×.

Fig. 328. Low-power view of a cross-sectioned renal papilla (man) whose most characteristic features at such a low magnification are the numerous uniform and regularly spaced lumina, each of which corresponds to a transverse section of a collecting tubule (compare with Fig. 336). H.E. staining. Magnification 24×.

Afferent arteriole   Macula densa

Fig. 329.   Proximal convoluted tubule

Macula densa

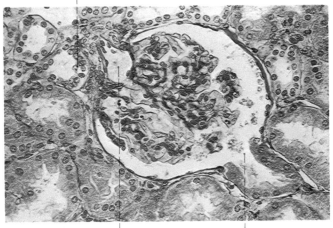

Fig. 330.   Vascular pole   Urinary pole

Distal convoluted tubule   Proximal convoluted tubule

Fig. 329. From an obliquely oriented interlobular artery (in the left half of the micrograph) originates an afferent arteriole that can be followed into the vascular pole of the renal corpuscle. Immediately above the corpuscle lies the cross section of a distal convoluted tubule with the macula densa adjacent to the Bowman's capsule (human kidney). Mallory-azan staining. Magnification 150×.

Fig. 330. Human renal corpuscle with urinary and vascular pole. At the former the parietal layer of the Bowman's capsule is continuous with the deeper staining and taller epithelium of the proximal convoluted tubule. The visceral layer of Bowman's capsule is transformed into the podocytes encircling the glomerular capillaries with their processes. Note prominent macula densa at the vascular pole. The tubular profiles surrounding each glomerulus belong to the proximal and distal convoluted tubules (for their differentiation see next micrograph). Mallory-azan staining. Magnification 240×.

Fig. 331. The prominent brush border found in the proximal convoluted tubules allows one to delineate these against the corresponding segments of the distal tubules that in addition have a less acidophilic epithelium. In some areas (e.g., midway at the upper margin of the micrograph) even the basal striations can be seen in the lining cells of the proximal convoluted tubules (cf. Fig. 19). Mallory-azan staining. Magnification 380×.

Fig. 332. A higher magnification of a medullary ray (tangential section through a human renal cortex, cf. Figs. 327, 333) clearly shows the differences between the proximal and distal segments of the uriniferous tubules. The taller and deeper staining (more acidophilic) epithelium of the proximal segments bulging into the lumen, together with its brush border, results in an irregular and poorly defined outline of the lumen. Contrary to this the lumina of the distal segments seem to be wider, showing a straight and clear-cut inner contour together with a proportionally lower epithelium. These differences become even more pronounced in the collecting tubules that show an increase of their inner and outer diameters together with a taller epithelial lining. H.E. staining. Magnification 240×.

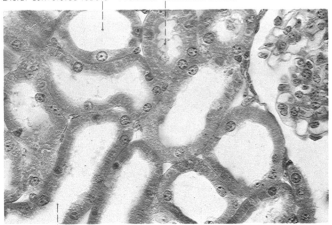

Proximal convoluted tubule   **Fig. 331.**

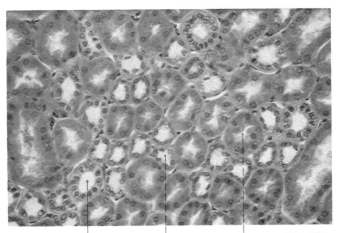

Collecting tubule   Distal tubule   Proximal tubule   **Fig. 332.**
(straight segment) (straight portion)

149

Urinary system – The renal tubules

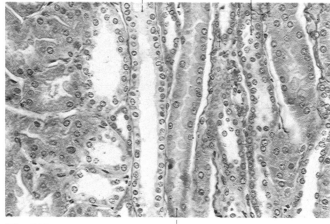

Collecting tubule   Thick limb of Henle's loop

Fig. 333.

Straight portion of proximal tubule

Collecting tubule

Fig. 334.

Bend of Henle's loop

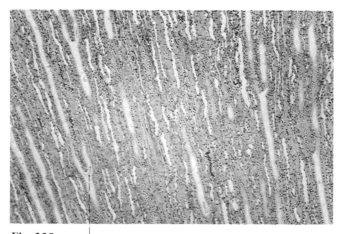

Fig. 335.

Fusion of collecting tubules

Fig. 333. Longitudinal section of a medullary ray (human kidney) with parts of the adjacent cortical labyrinth (at the left side of the micrograph). The straight portion of a proximal segment clearly displays its epithelial cells bulging into the lumen that hence is narrowed. The collecting tubule, however, lying to the left shows a wide lumen bordered by an even epithelial surface. The straight portion of the distal segment (lying to the right) is sectioned tangentially and therefore it can only be delineated from its proximal counterpart by the greater amount of nuclei it contains. Mallory-azan staining. Magnification 240×.

Fig. 334. Longitudinal section through the outer medullary zone (human kidney) in which the thin segments of Henle's loop lined by an extremely low epithelium are prominent. Their descending and ascending limbs join in a U-shaped apex that is always directed toward the medullary papillae. Mallory-azan staining. Magnification 150×.

Fig. 335. Longitudinal section through the outer medullary zone (human kidney) with numerous longitudinally sectioned collecting tubules which gradually merge to form finally the papillary ducts that open on the papillary apex. Note the numerous profiles of thin segments of Henle's loops lying between the collecting tubules to which they run parallel. Mallory-azan staining. Magnification 60×.

Fig. 336. In a cross-sectioned renal papilla (human kidney) the profiles of the collecting tubules are particularly prominent due to their large lumina and their high columnar epithelium (a low-power view is given in Fig. 328). The thin segments of Henle's loops and the blood capillaries running in parallel differ by (1) a slightly higher epithelium with its nuclei bulging into the lumen and by (2) a larger inner diameter of the former. The capillaries, however, show a smaller lumen and their nuclei are infrequently found in cross sections, but often their definite identification is facilitated because they contain red blood cells. H.E. staining. Magnification 240×.

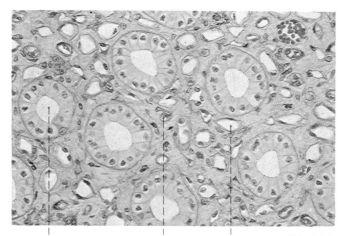

Collecting tubule   *Blood capillary   Thin segment
of Henle's loop*

Fig. 336.

Proximal convoluted tubule   *Efferent arteriole*

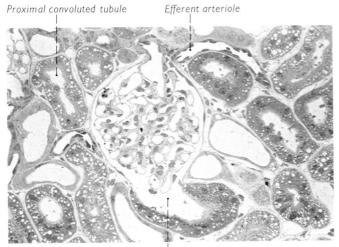

Fig. 337.

Urinary pole

Capsular space   *Brush border*

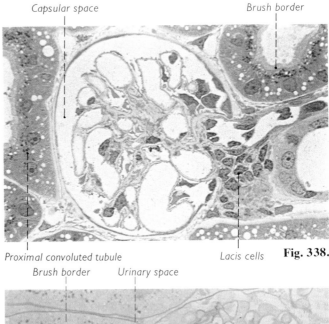

Proximal convoluted tubule   *Lacis cells*   **Fig. 338.**

Fig. 337. Semi-thin (ca. 1 μm thick) section through a renal corpuscle of a rat with vascular (on top) and urinary pole. The proximal convoluted tubules exhibit a prominent brush border at the base of which deeply stained minute granules can be identified that correspond to secondary lysosomes and residual bodies originating during the intracellular turnover of proteins. The large and translucent vacuoles possibly contain lipids (cf. Fig. 48a). Methylene blue-azur II staining. Magnification 240×.

Brush border   *Urinary space*

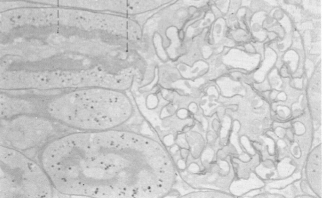

Fig. 338. Renal corpuscle (rat) with vascular pole and efferent arteriole clearly depicting a group of extraglomerular mesangial cells (also called lacis cells or cells of Goormaghtigh) that belong to the iuxtaglomerular complex and are wedged in between the afferent and efferent arteriole. Adjacent to the right side of the lacis cells parts of a macula densa can be seen, but the epitheloid iuxtaglomerular or "Polkissen" cells are not met by this semi-thin section. Methylene blue-azur II staining. Magnification 600×.

Fig. 339.

Renal capsule

Fig. 339. Renal corpuscle of a rat in a semi-thin section. The basal lamina of the glomerular capillaries together with those of the renal tubules is clearly depicted as a delicate violet band by means of a modified PAS technique. Note also the PAS-positive reaction of both the brush border in the proximal tubules and of the fine intracytoplasmic granules. Modified PAS staining. Magnification 480× (specimen courtesy of Prof. P. Böck).

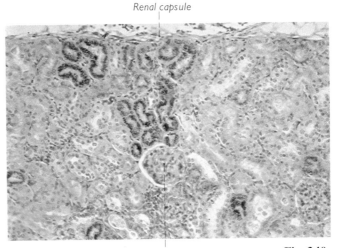

Fig. 340. Renal cortex of a rat with prominent "labeling" of the convoluted parts of the proximal tubules by the uptake and storage of the vital dye trypan blue. Nuclear fast red staining. Magnification 150×.

Glomerulus   **Fig. 340.**

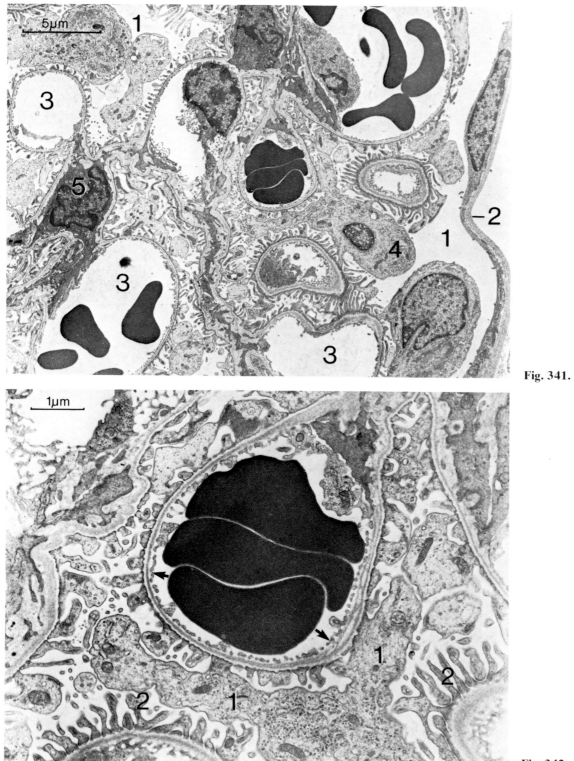

Fig. 341.

Fig. 342.

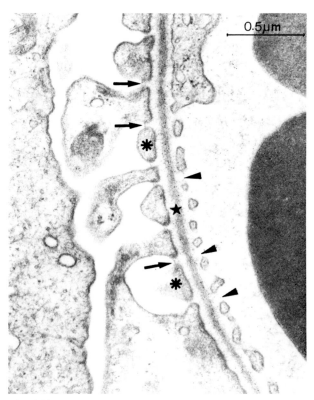

Fig. 343a.

Fig. 343b.

Fig. 343. a) Transverse section through a glomerular capillary to demonstrate the elements that constitute the filtration barrier. Due to the abundance of pores (▶) by which the endothelium is perforated, it usually appears as a row of cytoplasmic islets that are not interconnected by a membrane (diaphragm). The lamina densa (∗) of the basal lamina is particularly thick (50–60 nm) in the glomerulus and functions as a coarse filter for high molecular substances. The main filtration barrier, however, is represented by a delicate membrane, the slit membrane (→), that bridges the small clefts (slit pores) between adjacent foot processes (*). 1 = Erythrocyte. Magnification 40,000 ×.

b) A tangential section of a glomerular capillary better illustrates the sieve-like structure of the endothelium (1) and the close investment of the basal lamina (2) by innumerable interdigitating foot processes (3) of the podocytes that are separated by narrow (20–40 nm) intercellular clefts, the slit pores (→). Magnification 27,000×.

◀

Fig. 341. Lower-power electron micrograph illustrating parts of a rat renal corpuscle with its capsular space (1), the parietal layer of the capsule of Bowman (2) and several profiles of glomerular capillaries (3). Between the capillaries two kinds of cells can be identified: the perikarya of the podocytes (4) and the mesangial cells (5) that appear to be more electron dense. Magnification 4,000×.

Fig. 342. A higher magnification of the preceding micrograph clearly reveals the large number of pores (→) piercing the endothelium, and it illustrates the complex system of the podocyte processes which are subdivided into coarser primary (1) and slender secondary ones (2) that are called foot processes or pedicles. The primary processes (1) belong to the podocyte labeled "4" in the preceding micrograph. Magnification 14,000×.

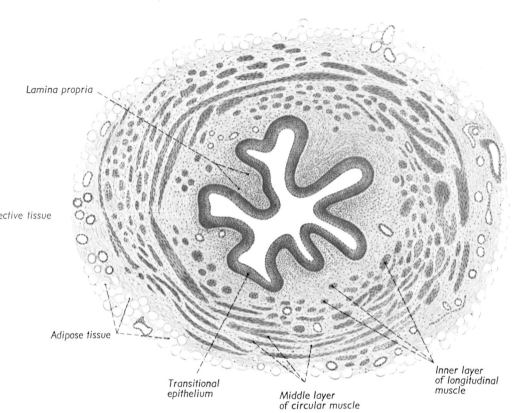

Lamina propria

Submucosal connective tissue

Adipose tissue

Transitional epithelium

Middle layer of circular muscle

Inner layer of longitudinal muscle

Fig. 344. Transverse section of a human ureter with its lumen being narrowed into a star-shaped outline due to the contraction of the muscularis (see also Table 15). This consists of an inner longitudinal, a middle circular and a less-developed outer longitudinal layer that are continuous with each other and hence poorly defined. This is explained by the assumption that the smooth muscle bundles form undisrupted strands wound around the long axis of the ureter in a spiral fashion (camera lucida drawing). H.E. staining. Magnification 30×.

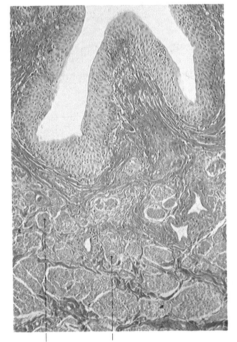

Bundles of smooth muscle cells

Fig. 345. Mucosal fold from a cross-sectioned human ureter with a thick transitional epithelium (its collapsed or contracted state), the underlying connective tissue (lamina propria) and the inner longitudinal muscle layer. One of the characteristics of the ureter is the loose arrangement of the smooth muscle bundles separated by an elaborate connective tissue framework. H.E. staining. Magnification 95×.

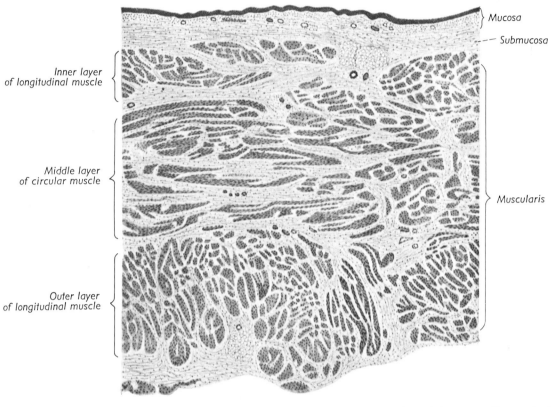

Mucosa

Submucosa

Inner layer
of longitudinal muscle

Middle layer
of circular muscle

Muscularis

Outer layer
of longitudinal muscle

Fig. 346. The human urinary bladder possesses a muscularis organized similarly to that of the ureter. But as the individual muscle bundles are parts of spirally wound continuous muscular strands, they are never oriented exactly circularly or longitudinally, but mostly more or less obliquely to the long axis of the bladder. The epithelium is rather thin due to a considerable degree of distension (camera lucida drawing). H.E. staining. Magnification 18×.

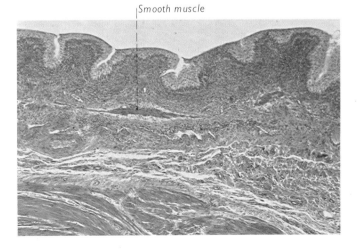

Smooth muscle

Fig. 347. Cross section of a moderately contracted urinary bladder (man). The mucosa is compressed into low irregular folds, and slender bundles of smooth muscle are visible beyond the highly cellular lamina propria. Then follows a loose connective tissue layer, the submucosa, that borders upon the innermost layer of the main muscular coat. H.E. staining. Magnification 60×.

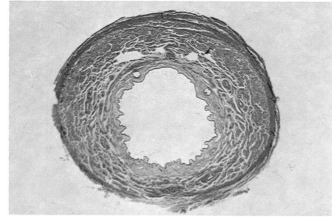

Fig. 348.

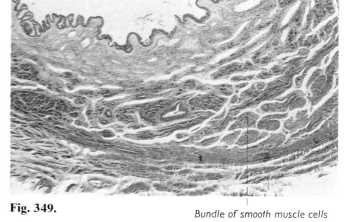

Fig. 349.   *Bundle of smooth muscle cells*

Artery

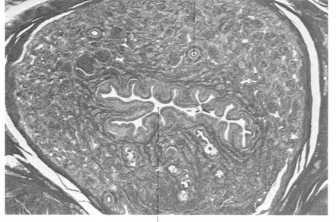

Fig. 350.   *Lumen of urethra*

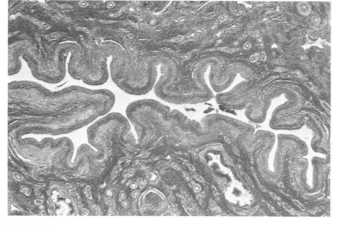

Fig. 351.

Fig. 348. A cross section of a female urethra shows its rather wide lumen and indefinite muscularis consisting of small muscle bundles but becoming more pronounced in its peripheral layers. Mallory-azan staining. Magnification 10×.

Fig. 349. A higher magnification discloses numerous veins lying in the lamina propria which aid in a firm closure of the urethra but no glands of Littré are included in this section. The epithelium is stratified columnar (for details cf. Fig. 66). Mallory-azan staining. Magnification 62×.

Fig. 350. Transverse section through the cavernous (spongy) portion of the male urethra. Its identification is made easy due to the surrounding mass of erectile tissue. H.E. staining. Magnification 10×.

Fig. 351. The higher magnification shows the epithelium to be two- or three-layered and belonging to the columnar variety according to the shapes of the cells comprising its surface layer. H.E. staining. Magnification 38×.

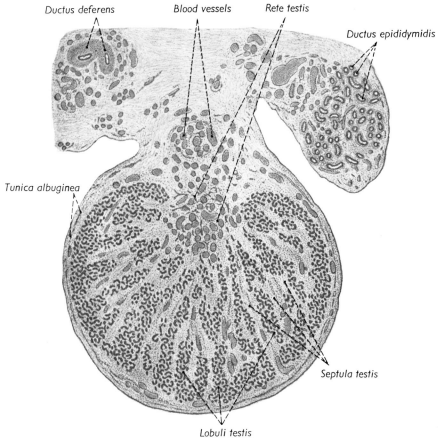

Ductus deferens   Blood vessels   Rete testis

Ductus epididymidis

Tunica albuginea

Septula testis

Lobuli testis

Fig. 352. Low-power view of an immature human testis of an infant. Radiating from the capsule (tunica albuginea) toward the hilus are connective tissue septa (septula testis) that subdivide the organ into lobules, each of which regularly contains several of the intricately coiled tubules, the convoluted seminiferous tubules. These empty via the rete testis located within the mediastinum into the ductuli efferentes passing into the head of the epididymis (camera lucida drawing). H.E. staining. Magnification 16×.

Primordial germ cell

Fig. 353. The seminiferous tubules of this human fetal testis are predominantly solid (germinal) cords consisting of only two types of cells, the prevailing type being the primitive Sertoli cells, which clearly stand out by their closely spaced and deeply staining nuclei. The second cell type consists of the primordial germ cells, which migrate along the dorsal mesentery of the hindgut into the gonadal ridge. They can be identified by (1) their large size, (2) their light-staining cytoplasm and (3) their spherical nuclei. Mallory-azan staining. Magnification 380×.

Fig. 353.   Primordial germ cell

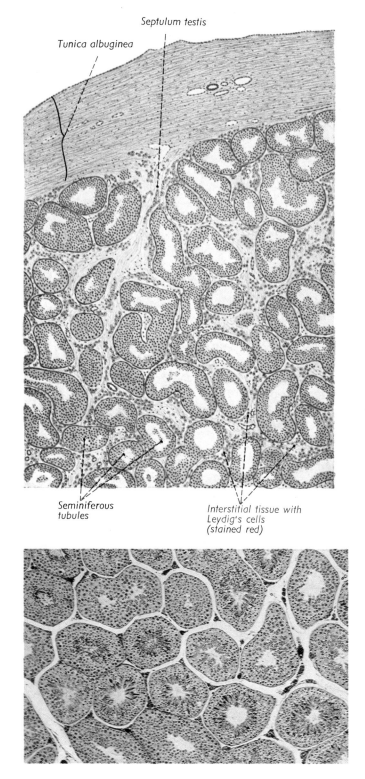

Tunica albuginea

Septulum testis

Seminiferous tubules

Interstitial tissue with Leydig's cells (stained red)

Fig. 354. Section from the peripheral parts of a mature human testis that is enclosed in a dense fibrous capsule, the tunica albuginea, whose surface is covered by the visceral layer (epiorchium) of the tunica vaginalis. In the interstices between the seminiferous tubules the loosely aggregated and stronger acidophilic interstitial cells of Leydig can be seen (camera lucida drawing). H.E. staining. Magnification 40×.

Fig. 355. Low-power view of several seminiferous tubules from a rat testis. Such specimens are often used in histological courses because in rodents the spermatogenic activity occurs in a wave-like fashion along the tubules. Therefore in each of the individual tubules cross-sectioned at different levels definite stages of spermatogenesis will prevail and hence can be seen particularly clearly. Weigert's iron-hematoxylin and benzo light bordeaux staining. Magnification 60×.

Fig. 356. Cross-sectioned seminiferous convoluted tubule from human testis showing nearly all stages of spermatogenesis. The cells directly adjacent to the basal lamina and fitted with a spherical nucleus are the spermatogonia. On their inner side lie the primary spermatocytes that are larger cells but also equipped with spherical nuclei. They originate from the former by a mitotic cell division followed by a growth period. These give rise to two secondary spermatocytes (= prespermatids) that are half the size of their mother cells and it is the first maturation division by which they are formed. Hardly visible in this specimen are the spermatids that result from the second maturation division while the comma-shaped heads of the spermatozoa stand out clearly as intensely stained corpuscles. In the connective tissue interstices groups of interstitial (Leydig) cells of which the one at the upper right side of the micrograph contains the proteinaceous crystalloids of Reinke (for details see Fig. 38). Mallory-azan staining. Magnification 240 × .

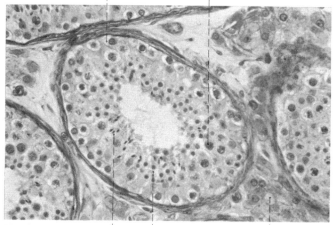

Spermatogonium   Primary spermatocyte

Fig. 356.   Spermatozoa   Secondary spermatocyte   Interstitial cells

Primary spermatocyte   Secondary spermatocyte

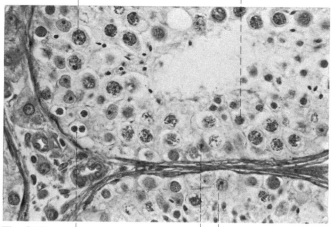

Fig. 357. A higher magnification of the epithelial lining of a seminiferous tubule (same as the foregoing specimen) discloses the nuclei of the Sertoli cells. They differ from all the various germ cells by regularly showing an indistinct chromatin but a prominent nucleolus. Note telophase of a spermatogonial division at the lower left margin of the tubule. Mallory-azan staining. Magnification 380 × .

Fig. 357. Telophase of spermatogonium Nuclei of Sertoli cells

Secondary spermatocyte

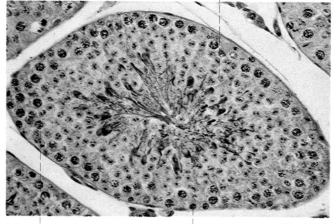

Fig. 358. Cross-sectioned seminiferous tubule from a rat testis. Irrespective of a few scattered spermatogonia the epithelium is governed by primary spermatocytes (to be recognized by their prominent chromatin) and their daughter cells, the secondary spermatocytes. The lumen of the tubule is crowded with tails of spermatozoa whose heads seem to lie deep between the germ cells. Actually small groups of them (mostly consisting of eight) are enveloped by cytoplasmic processes of the Sertoli cells. Weigert's iron-hematoxylin and benzo light bordeaux staining. Magnification 240 × .

Primary spermatocyte   Spermatogonium   **Fig. 358.**

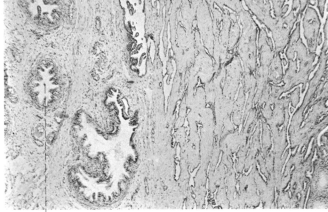

Ductulus efferens

Fig. 359.

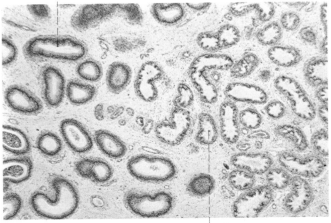

Ductus epididymidis

Fig. 360.

Ductulus efferens

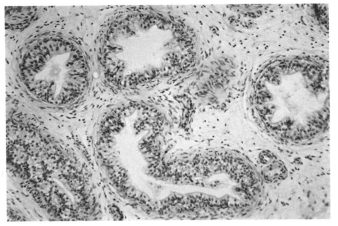

Fig. 361.

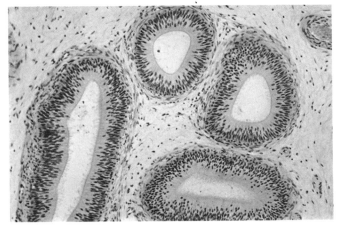

Fig. 362.

Fig. 359. Section through the testicular mediastinum (right side) and the adjacent head (caput) of the human epididymis. The richly branching, narrow and anastomosing tubules of the rete testis constitute the beginnings of the excretory ducts of the testis. They empty into the ductuli efferentes located within the head of the epididymis. H.E. staining. Magnification 38×.

Fig. 360. Low-power view of a human epididymis illustrating both types of ductules occurring in this organ. The tortuous ductuli efferentes are located in the head (right side of the micrograph) while the even more convoluted ductus epididymidis occupies the body (corpus) and the tail (cauda) of the epidydimis. The lumen of the ductuli efferentes has a characteristic serrated outline by which these tubules can be readily distinguished from the ductus epididymidis that regularly shows a straight inner contour. H.E. staining. Magnification 25×.

Fig. 361. With a higher magnification it becomes evident that both the height of the epithelial cells and the number of their layers alternate at quite regular intervals around the entire circumference of a ductulus efferens. The parts projecting into the lumen are made of a pseudostratified or stratified tall columnar epithelium furnished with kinocilia, while the crypt-like depressions are lined by a nonciliated pseudostratified cuboidal epithelium. The kinocilia are invisible in this preparation because the magnification is too low. H.E. staining. Magnification 96×.

Fig. 362. In contrast the pseudostratified, tall columnar epithelium of the ductus epididymidis is very uniform in shape and is fitted with stereocilia. These specializations of the epithelial surface are invisible because of the low magnification (for details see Fig. 75). The lumina are often crowded with spermatozoa. H.E. staining. Magnification 96×.

Artery within pampiniform plexus

Cremaster muscle   Ductus deferens   **Fig. 364.**

Fig. 363. Cross section through a human spermatic cord containing the ductus deferens together with numerous blood vessels, most of which belong to the venous pampiniform plexus. As these veins have an extremely thick three-layered media they easily can be confused with arteries. The striated cremaster muscle is seen in the lower left corner of this micrograph. H.E. staining (faded). Magnification 9.5 ×.

Fig. 364. Transverse section through the ampulla of the human ductus deferens. In contrast to the ductus deferens itself here the epithelium covers a complicated network of anastomosing folds and the three layers of the muscularis are less distinct with its circular components prevailing. The ampulla of the ductus deferens can be distinguished from the seminal vesicles (with which it is sometimes confused) by (1) its narrow lumen, (2) its thicker muscularis and (3) its less welldeveloped mucosal folds (cf. Fig. 367). H.E. staining. Magnification 17×.

Fig. 365. The human ductus deferens is characterized especially by its extremely thick muscularis subdivided into three definite layers. These are assumed to be parts of undisrupted muscular strands that encircle the ductus in a spiral fashion. As these run steep at the outside, nearly circular in the middle and steep again at the inside, a cross section must give the impression of an outer longitudinal, a middle circular and an inner longitudinal muscle layer. The epithelium is pseudostratified columnar and is fitted with stereocilia that vanish at the end of the ductus deferens. As found in other comparable tubes, e.g., the ureter, the lumen appears starshaped (cf. Table 15) due to the contraction of the muscularis (camera lucida drawing). H.E. staining. Magnification 53 ×.

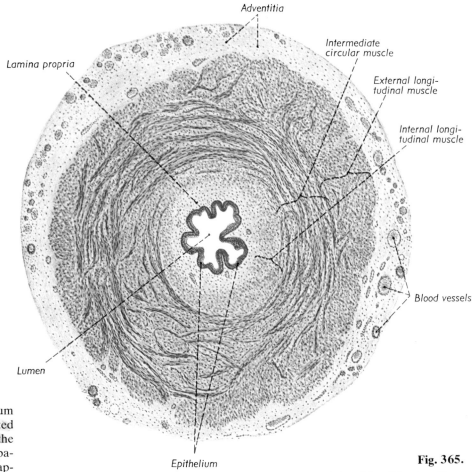

Adventitia

Intermediate circular muscle

External longitudinal muscle

Lamina propria

Internal longitudinal muscle

Blood vessels

Lumen

Epithelium   **Fig. 365.**

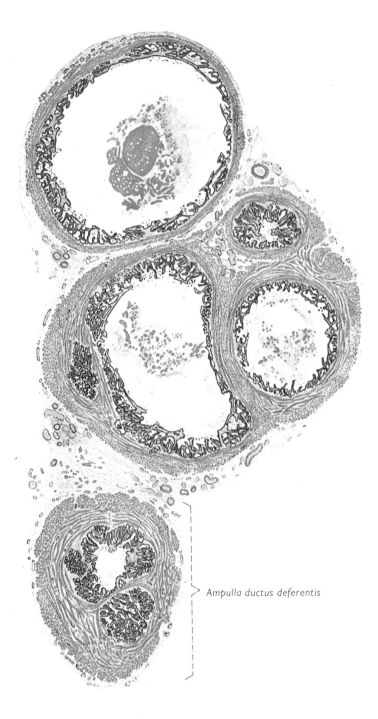

Ampulla ductus deferentis

Fig. 366. Low-power view of a seminal vesicle together with the ampulla of the ductus deferens that can be distinguished one from another by the size of their lumina and the width of their muscle coats (camera lucida drawing). Compare also with the micrographs of Figs. 364 and 367. H.E. staining. Magnification 10 ×.

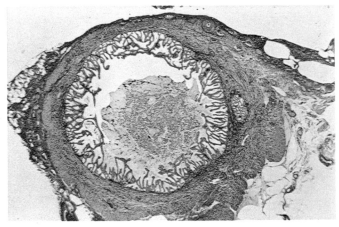

Fig. 367.

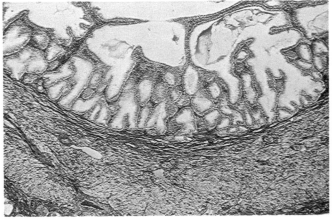

Fig. 368.

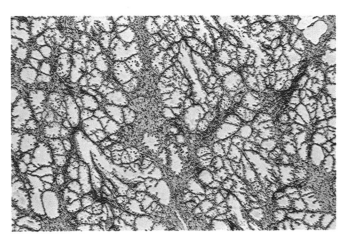

Fig. 369.

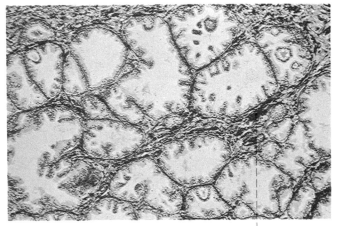

Fig. 370.   *Smooth muscle*

Fig. 367. Cross section of a human seminal vesicle with its mucosa thrown into a characteristic complex pattern of interconnected folds. The muscularis mainly consists of obliquely and longitudinally oriented fiber bundles that are not arranged in definite layers. The differentiation against the ampulla of the ductus deferens is based on (1) the wider lumen, (2) the much more elaborated mucosal folds and (3) the considerably lesser muscularis. Mallory-azan staining. Magnification 17×.

Fig. 368. The higher magnification shows the filigree-like texture of the mucosa due to its folds anastomosing frequently with each other. As already described for the gall bladder (cf. Fig. 286) also in this mucosa many irregular chambers lined with an epithelium can be found. Their greater number together with the much thicker muscularis allow for a clear distinction when compared with the gall bladder. Mallory-azan staining. Magnification 48×.

Fig. 369. Low-power view of the human prostate gland. A section through this tubulo-alveolar gland shows rather large irregularly shaped and frequently indented cavities between which an excretory duct is only found in rare exceptions. This last feature allows for a clear differentiation from the lactating mammary gland with which it is often confused (for differential diagnosis cf. Figs. 373–376 and Table 14). Mallory-azan staining. Magnification 38×.

Fig. 370. At higher magnification it is seen that the columnar epithelium of the alveoli varies in height and is thrown into delicate folds that provide the secretory portions with a frill-like inner contour. A unique (!) feature of the prostate gland is a vast number of interlacing smooth muscle bundles coursing within the connective tissue septa. Mallory-azan staining. Magnification 150×.

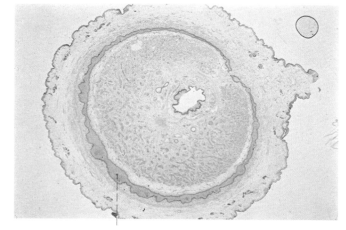

Inner epithelium of prepuce

Fig. 371. Cross section through the glans penis of an infant at the level of the fossa navicularis. As the inner epithelial lining of the prepuce is still "glued" to the outer epithelium covering the glans, it is encircled by a solid epithelial glando-preputial lamella. In the vicinity of its external orifice the urethra is lined by a stratified noncornified squamous epithelium. H.E. staining (faded). Magnification 10×.

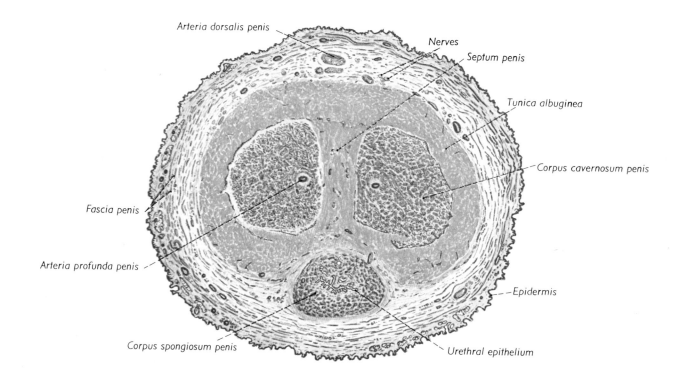

Arteria dorsalis penis

Nerves

Septum penis

Tunica albuginea

Corpus cavernosum penis

Fascia penis

Arteria profunda penis

Corpus spongiosum penis

Urethral epithelium

Epidermis

Fig. 372. Cross section through the shaft of a human penis (camera lucida drawing). For details of its corpus spongiosum and the urethra it encloses see Fig. 350 and 351. H.E. staining. Magnification 4×.

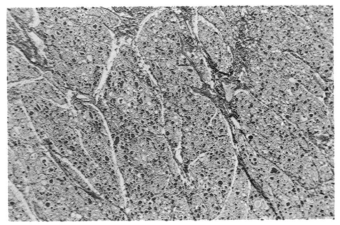

Fig. 373.

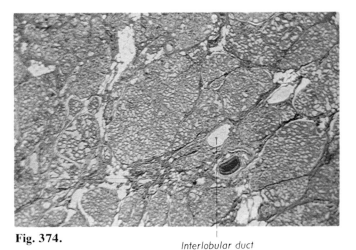

Fig. 374.

Interlobular duct

Bronchus

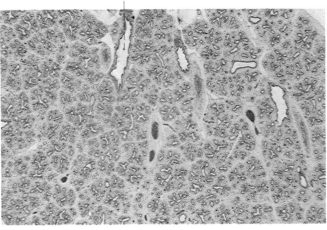

Fig. 375.

Figs. 373.–376. When comparing sections of those alveolar glands that are often confused with each other like the thyroid (Fig. 373), prostate (Fig. 376) and lactating mammary gland (Fig. 374) the prostate gland clearly differs from the two others by its lack of distinct lobular subdivisions. Furthermore it is characterized by a unique feature, i.e., the smooth muscle cells found in the connective tissue interstices. The lactating mammary gland is distinguished from the thyroid by its consistently large excretory ducts, whereas the latter is marked by secretory units (follicles) that vary considerably in size and in the amount of colloid they contain. In this context also the fetal lung (Fig. 375) has to be mentioned because, like the glands, it originates as an epithelial sprout and hence shows a similar growth modality. The fetal lung is most often confused with an active mammary gland, particularly when the preparation has not been first examined with the lowest power objective, lest, its most typical feature, i.e., the bronchial primordia, be overlooked. These can be identified by the hyaline cartilage found in their walls (cf. also Fig. 121 and Table 14). A further characteristic is the highly cellular and very loose connective tissue that thereby discloses its mesenchymal nature. All figures: Mallory-azan staining. Magnification 38×.

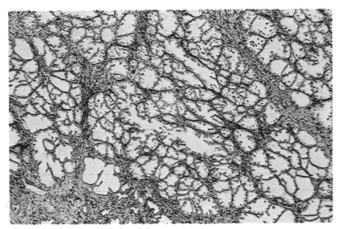

Fig. 376.

165

Fig. 377.

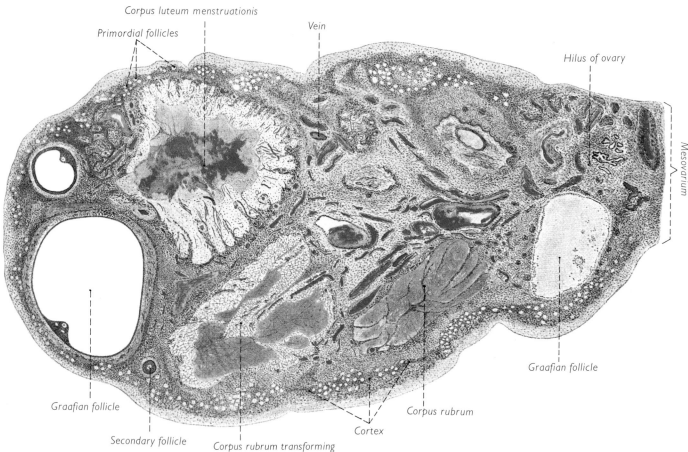

Corpus luteum menstruationis

Primordial follicles

Vein

Hilus of ovary

Mesovarium

Graafian follicle

Graafian follicle

Secondary follicle

Corpus rubrum transforming into a corpus albicans

Cortex

Corpus rubrum

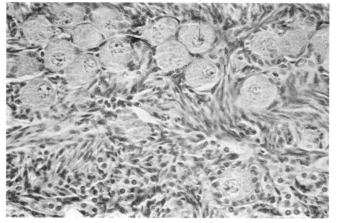

Nucleus of oocyte

Fig. 378.

Fig. 377. Complete transverse section (because the mesovarium is included) of a human ovary. Slightly schematized drawing composed from a series of histologic specimens (after Patzelt: *Histologie,* 3rd ed., 1948).
In order to identify the various developmental stages of the ovarian follicles, suitable areas must be selected by using the low-power objective. As this can be difficult in human specimens, ovaries from laboratory animals are often shown in histology courses. H.E. staining. Magnification 10×.

Fig. 378. Several primordial follicles in the cortex of a feline ovary. These consist of a primary oocyte enveloped by a single layer of a squamous to cuboidal (follicular) epithelium. Primordial follicles are occasionally confused with spinal ganglion cells (cf. Figs. 11, 172, 471), particularly if the section is merely viewed with a high-power objective. H.E. staining. Magnification 240×.

Primary follicles

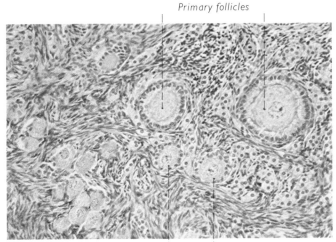

Fig. 379.

Primordial follicles

Membrana granulosa

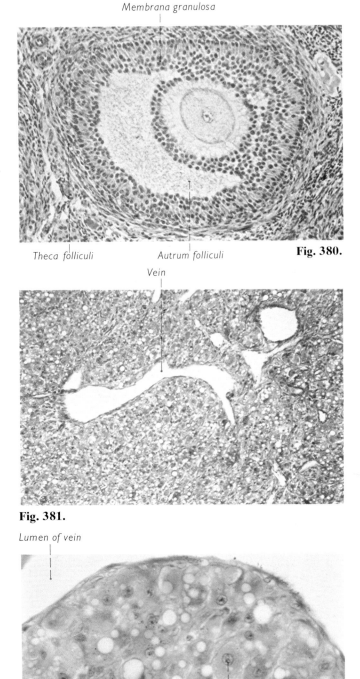

Theca folliculi   Autrum folliculi   **Fig. 380.**

Vein

Fig. 379. Cortical zone of a human ovary showing several primordial and two primary follicles. The latter are characterized (1) by a higher and in later stages multilayered epithelium consisting of cuboidal or low columnar "granulosa" cells, (2) by a hyaline membrane (zona pellucida) interposed between the oocyte and the innermost layer of the granulosa cells and (3) by the large size of the oocyte that has grown and matured to a secondary oocyte. H.E. staining. Magnification 150×.

Fig. 380. Secondary (antral) follicle with a crescentic cavity (antrum folliculi) and a developing cumulus oophorus. The multilayered follicular epithelium constitutes the wall of this globule that becomes successively filled with fluid (liquor folliculi). The connective tissue immediately adjacent to the follicle has been transformed into cellular strands encircling the follicle as the so-called theca folliculi, which, however, has not yet differentiated into the later theca interna and externa. H.E. staining. Magnification 150×.

Fig. 381.

Lumen of vein

Fig. 381. Low-power view of the central parts of a feline corpus luteum with several larger veins. The lipid droplets within the granulosa lutein cells are dissolved in routine histological preparations, and hence these cells appear highly vacuolated. Mallory-azan staining. Magnification 96×.

Fig. 382. A higher magnification from the center of the preceding micrograph better illustrates the large size of the granulosa lutein cells and their variant states of vacuolization. Mallory-azan staining. Magnification 380×.

Fig. 382.

Nucleus of granulosa lutein cell

167

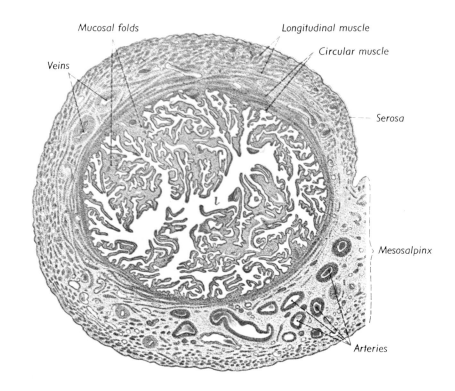

Fig. 383. Complete transverse section through a human uterine (Fallopian) tube at the level of its ampulla (camera lucida drawing). Peculiar characteristics are the delicate and highly branched mucosal folds together with a muscularis not subdivided into distinct layers. If the specimen is obtained intact and fixed very carefully, its outer serosal covering will be preserved (but it is often missing). For differential diagnosis see Table 15. H.E. staining. Magnification 22×.

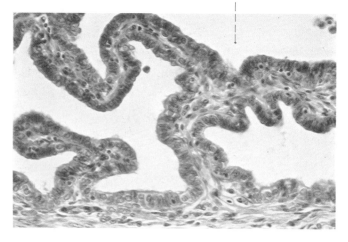

Fig. 384. Human oviductal mucosa and its folds under higher magnification. The simple columnar epithelium is partly furnished with kinocilia with the amount of ciliated cells undergoing cyclic changes (cf. Figs. 63 and 74). The subepithelial lamina propria consists of reticular connective tissue. H.E. staining. Magnification 95×.

Fig. 385.

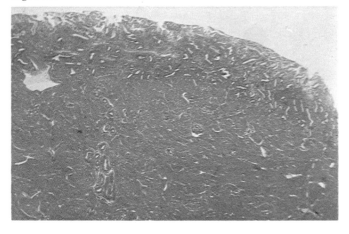

Correlative illustration of some typical appearances of the uterine mucosa (endometrium) as they regularly occur in every menstrual cycle.

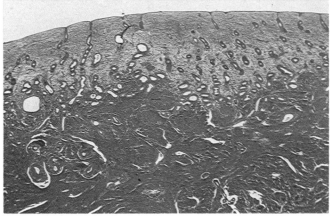

Fig. 386.

Fig. 385. The menstrual stage, ranging from the 1st to the 4th day after onset of menstruation, results in a restoration of the surface epithelium after the "functionalis" has been completely discarded. Regeneration originates from the blind ends of the endometrial glands that always remain in the "basalis" (human uterus, 2nd day of menstruation). H.E. staining. Magnification 17×.

Fig. 386. Under the influence of the ovarian estrogens, particularly the upper portions of the endometrium (= functionalis) increase in height during the proliferative stage (from 5th to 14th day of menstrual cycle) while the "basalis" (approx. 1 mm thick) is only moderately involved in this growth period. On the other hand, the latter is not discarded during menstruation. The endometrial glands appear in this phase as straight tubules (approx. 12th day of cycle). H.E. staining. Magnification 17×.

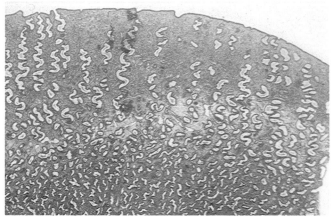

Fig. 387.

Fig. 387. At the end of the secretory stage (from 15th to 28th day of cycle) the tubular glands are highly tortuous and hence present a serrated outline in a section. As the upper portion of the functionalis contains not only more cells but additional connective tissue elements transformed into large "pseudodecidual" cells, it appears to be "dense" and therefore is known as "compacta." In contrast to this, the deeper and highly glandular mucosal layers are defined as "spongiosa" (endometrium from the 26th day of the cycle). H.E. staining. Magnification 17×.

Fig. 388. At a higher magnification (detail from the upper left corner of Fig. 387) the simple columnar epithelium is shown to be devoid of kinocilia. Note the large amount of cells lying in the "compacta" of the uterine mucosa. H.E. staining. Magnification 120×.

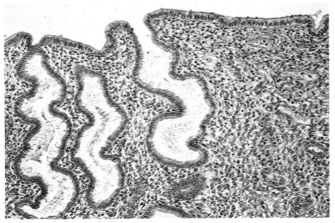

Fig. 388.

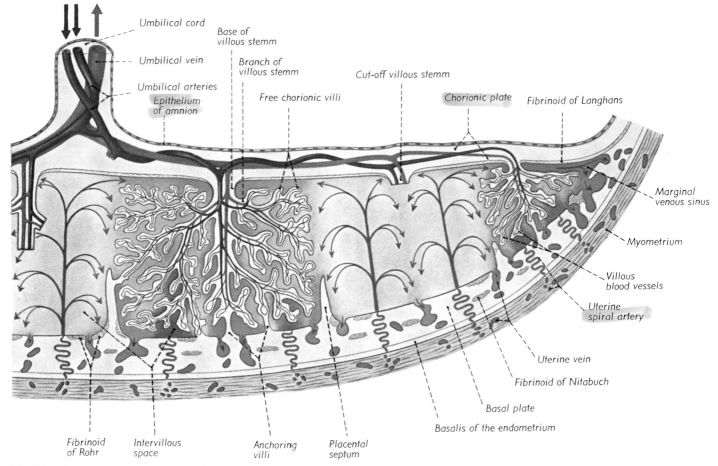

Umbilical cord
Base of
villous stemm
Umbilical vein
Branch of
villous stemm
Cut-off villous stemm
Umbilical arteries
Epithelium
of amnion
Free chorionic villi
Chorionic plate
Fibrinoid of Langhans
Marginal
venous sinus
Myometrium
Villous
blood vessels
Uterine
spiral artery
Uterine vein
Fibrinoid of Nitabuch
Basal plate
Basalis of the endometrium
Fibrinoid
of Rohr
Intervillous
space
Anchoring
villi
Placental
septum

Fig. 389. Schematic presentation of the placental circulation (modified from von Heidegger and Starck). The blood enters the intervillous spaces via the spiral arteries traversing the basal plate and then shoots upward onto the chorionic plate due to the high pressure it is under. Reflected at the chorionic plate it falls back and then circulates through the labyrinthic intervillous spaces, and is finally drained into the uterine veins.

Fig. 390. In a low-power view of a complete cross section of a human placenta the different components of this complex organ can be best identified as follows. Its fetal portion consists of (1) the chorionic plate covered on its surfaces by either the amniotic or the chorionic epithelium, (2) the highly branching villous stems (= cotyledons) originating from the chorionic plate that are partially anchored to the opposite maternal portion by means of "anchoring" villi (cf. Fig. 391).
The maternal portion is made of (1) the basal plate that is a derivative of the basal decidua and (2) its septal projections (= placental septa) that provide an incomplete separation between the individual cotyledons (camera lucida drawing). H.E. staining. Magnification 27.5×.

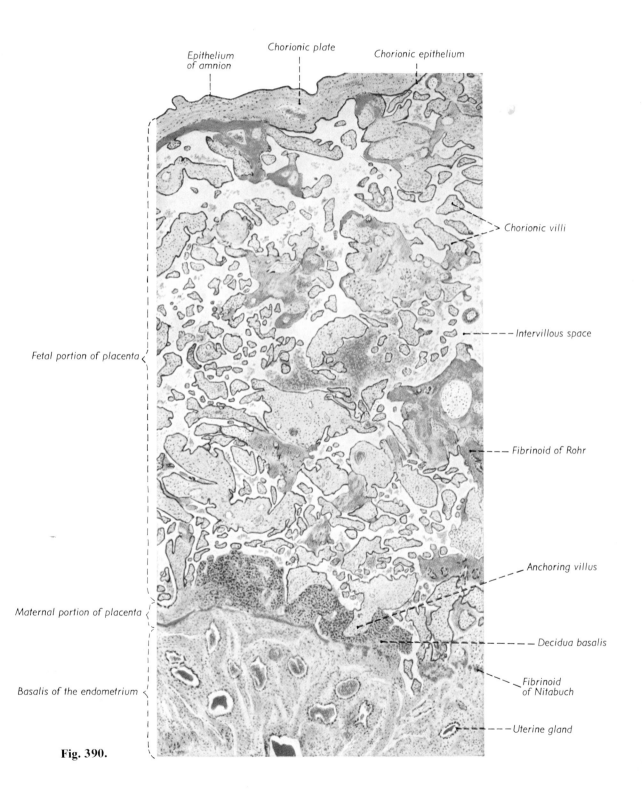

Epithelium
of amnion

Chorionic plate

Chorionic epithelium

Chorionic villi

Fetal portion of placenta

Intervillous space

Fibrinoid of Rohr

Anchoring villus

Maternal portion of placenta

Decidua basalis

Fibrinoid
of Nitabuch

Basalis of the endometrium

Uterine gland

Fig. 390.

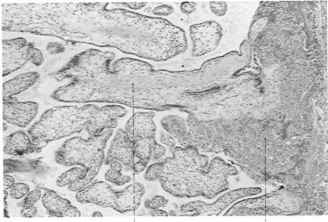

Fig. 391.   *Anchoring villus*   *Decidual cells*

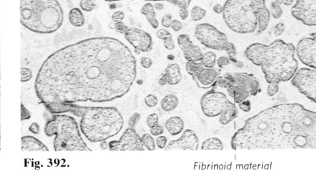

Fig. 392.   *Fibrinoid material*

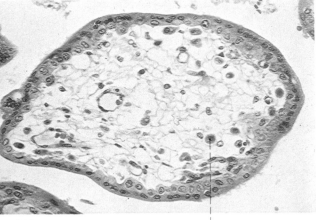

Fig. 393.   *Hofbauer's cell*

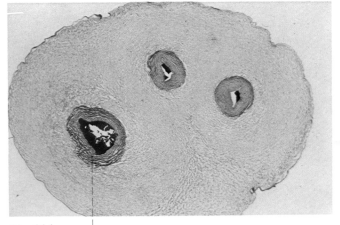

Fig. 394.   *Umbilical vein*

Fig. 391. Fusion site of an anchoring villus with the basal plate (mature human placenta). The deeper staining orange strands coursing between the decidual cells are known as Nitabuch's fibrinoid (cf. also Fig. 390). H. and chromotrop staining. Magnification 60×.

Fig. 392. Several cross-sectioned placental villi of different sizes from a mature human placenta. Note both the numerous vascular lumina seen within the connective tissue of the villi and the deposits of fibrinoid (stained orange) in the intervillous spaces. H. and chromotrop staining. Magnification 60×.

Fig. 393. Cross-sectioned villus from an early human placenta (about fourth month; fetus: 10 cm C.R. length) covered by a double-layered epithelium. The surface is lined by what is known as the syncytiotrophoblast. Its cells originate from the subjacent cytotrophoblast and later fuse with each other. The large and intensely stained cells found in the connective tissue core are the Hofbauer cells that are closely related to histiocytes. H.E. staining. Magnification 240×.

Fig. 394. Transverse section of a human umbilical cord obtained at delivery. Its surface is covered by the simple amniotic epithelium, and embedded in its mucous connective tissue (= Wharton's jelly) are seen the two umbilical arteries and a single vein. As usual the three blood vessels are found in an extremely contracted state after birth. No remnant of the allantoic duct is visible in this specimen. Mallory-azan staining. Magnification 10×.

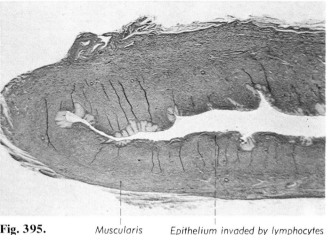

Fig. 395. Transverse section through a human vagina that has been split longitudinally into two identical halves to illustrate the composition of its wall. The stratified and noncornified squamous epithelium (for details cf. Fig. 64) is partially infiltrated by lymphatic aggregations. Its thick connective tissue lamina propria is regularly free of any glands but is rich in blood vessels, predominantly venous plexuses. The muscularis consists of interlacing bundles of smooth muscle cells (for further identifying characteristics see Fig. 269). Mallory-azan staining. Magnification 7 ×.

Fig. 395.   *Muscularis*   *Epithelium invaded by lymphocytes*

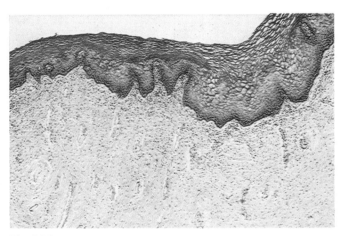

Fig. 396. The vaginal epithelium is particularly rich in glycogen that can be stained selectively, e.g., a deep red with Best's carmine as shown in this preparation. Together with the desquamated epithelial cells it is transferred into the vaginal lumen where it is metabolized to lactic acid by the Döderlein's bacilli. H. and Best's carmine staining. Magnification 60 ×.

Fig. 396.

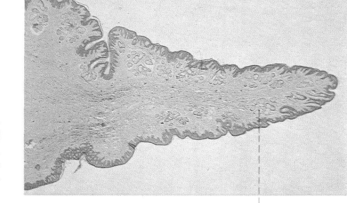

Fig. 397. Cross-sectioned human minor labium which in contrast to its major counterpart is regularly free of any hairs and sweat glands but is rich in sebaceous glands not connected with hairs. The stratified squamous epithelium is only slightly cornified and its basal cells are pigmented. H.E. staining. Magnification 8 ×.

Fig. 397.   *Sebaceous gland*

173

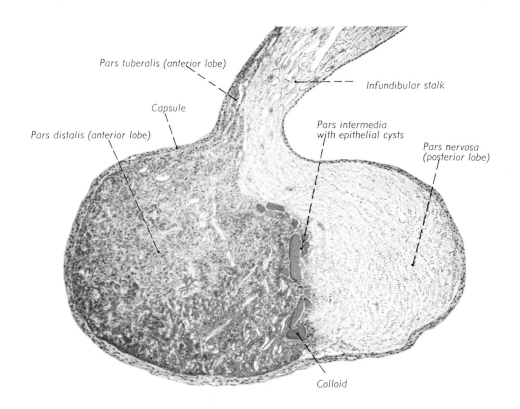

Pars tuberalis (anterior lobe)

Infundibular stalk

Capsule

Pars intermedia
with epithelial cysts

Pars distalis (anterior lobe)

Pars nervosa
(posterior lobe)

Colloid

Fig. 398. Low-power view of a complete midsagittal section through a human hypophysis to demonstrate its different components (for details see the labeling of the figure). Missing from this preparation is the posterior part of the pars tuberalis which covers the dorsal aspect of the infundibular stalk. Even at this very low magnification, the different staining reactions reflecting the uneven distribution of the various cell types within the sectional area of the anterior lobe are prominent (camera lucida drawing). H.E. staining. Magnification 10×.

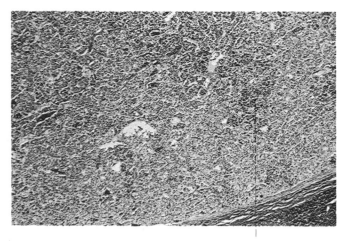

Group of basophils

Fig. 399. Another low-power view at a slightly increased magnification also shows the uneven distribution of the different cell types within this sectional area: subjacent to the capsule (at the lower margin of the micrograph) predominantly chromophobe cells are seen, while toward the lobular center the acidophils and groups made of basophils prevail. Hence to find the various cell types one has to search for the appropriate area with a low-power objective. Furthermore many ordinary preparations are of a poor quality due to the difficulties in obtaining well-preserved fresh human specimens. Materials obtained at autopsies (most of them not performed earlier than 24 hours after death) already show numerous postmortem artifacts. Mallory-azan staining. Magnification 38×.

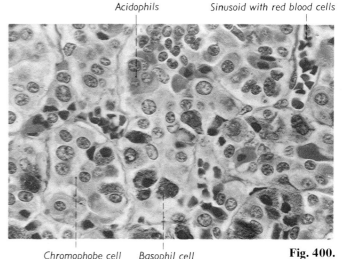

Acidophils   Sinusoid with red blood cells

Chromophobe cell   Basophil cell   **Fig. 400.**

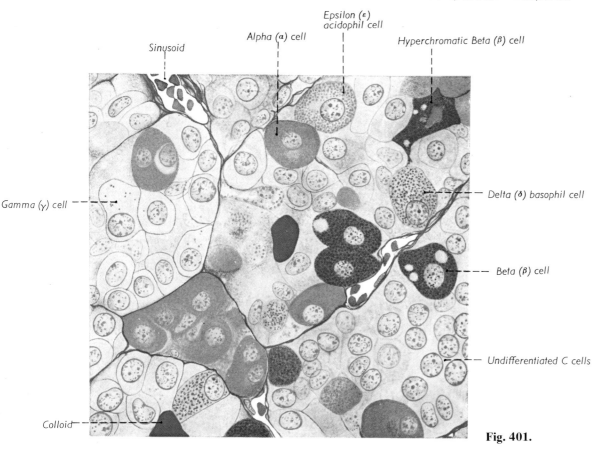

Sinusoid

Alpha (α) cell

Epsilon (ε) acidophil cell

Hyperchromatic Beta (β) cell

Gamma (γ) cell

Delta (δ) basophil cell

Beta (β) cell

Undifferentiated C cells

Colloid

Fig. 401.

Figs. 400 and **401.** While a camera lucida drawing allows one to sketch the different cell types of the anterior hypophyseal lobe from different human specimens and thereby combine them into a slightly schematized and idealistic picture (Fig. 401), this grouping rarely is found in an original section (Fig. 400). But by comparing both figures, a nearly complete identification of the different cell types can be achieved. Note, however, that in this specimen the acidophils show a stain that is more orange than the red usually seen. Stainings: Mallory-azan (Fig. 400) and a special modification called "Kresazan" (Fig. 401). Magnification 480× (Fig. 400) and 980× (Fig. 401).

175

Infundibular stalk   Pars tuberalis   Capsule

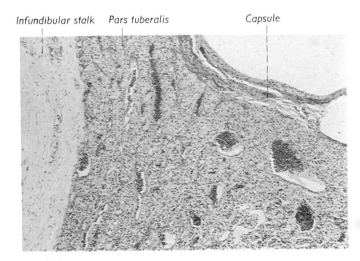

Fig. 402.

Fig. 402. The pars tuberalis with its adjacent infundibular stalk from a midsagittal section of a human hypophysis. Note that contrary to the anterior lobe the pars tuberalis consists of uniform cells arranged into cords with numerous blood vessels between. These show rather wide lumina and are part of the hypophyseal portal system. Mallory-azan staining. Magnification 60×.

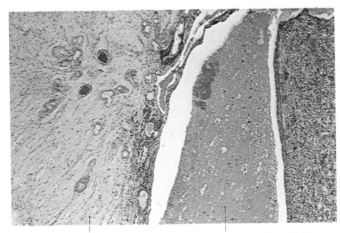

Fig. 403.   *Neural lobe*   *Colloid in remnant of vestigial space*

Fig. 403. Low-power view of the pars intermedia of a human hypophysis. This is predominantly occupied by a large cyst whose content (= colloid) is separated from its epithelial wall by a broad cleft due to the withdrawal of water during the embedding procedure. On the right side of the micrograph the adjoining portions of the anterior lobe can be seen while to the left the remnants of the intermediate lobe blend with the neural part of the gland. The former consist of small strands of cells and epithelial vesicles that closely resemble secretory units. Mallory-azan staining. Magnification 38×.

Colloid in epithelial cyst

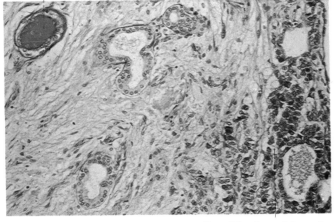

Fig. 404.   *Basophils of pars intermedia*

Fig. 404. A higher magnification of a small area located in the middle third of the foregoing micrograph discloses to a better advantage not only the epithelial cysts, one of which contains colloid, but also the highly basophilic cells growing out from the pars intermedia into the neural lobe. Other details regarding the cellular organization and the fiber architecture of the neural lobe can only be seen with special staining procedures. Mallory-azan staining. Magnification 150×.

Fig. 405. Low-power view of a complete sagittal section through the human pineal body (epiphysis cerebri). Though not found in every histology course due to the difficulty obtaining well-preserved specimens, the epiphysis cerebri is often confused with the parathyroid, particularly when the specimen is not thoroughly studied with a low-power objective. Differential features are the following: (1) the great difference in size (the parathyroid is considerably smaller, cf. Fig. 411), (2) the poor staining reactions of the epiphysis due to its high amount of nerve fibers and its faintly staining cells and (3) more prominent connective tissue septa that clearly separate the epiphyseal cells into lobules. The unique feature of sand granules (acervulus) occurring in the pineal body is missing here and there and is more often overlooked, if the slide is not studied at first with a low-power objective. It also can be seen in this section as a granular material stained a dark violet and consisting of mulberry-shaped and concentrically layered calcareous concretions. H.E. staining. Magnification 10×.

Sand granules (acervulus)

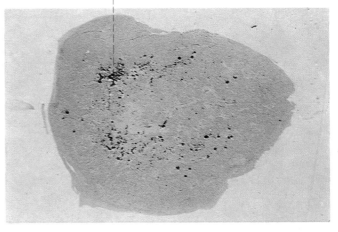

Fig. 405.

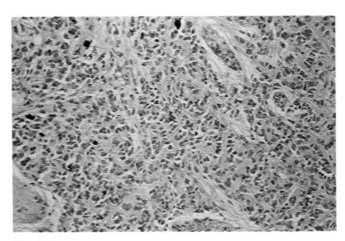

Fig. 406.

Figs. 406 and **407.** In ordinary preparations even a higher magnification only shows the cells being arranged into larger groups without allowing for a differentiation into specific pinealocytes and glial cells. Contrary to the parathyroidal cells the cellular elements of the epiphysis are never closely apposed polyhedral structures and hence epithelioid in nature (cf. Fig. 412). In the middle of Fig. 407 two cross-sectioned capillaries containing an erythrocyte can be seen. H.E. staining. Magnifications 150× (Fig. 406) and 380× (Fig. 407).

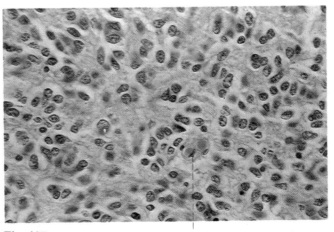

Fig. 407.

Blood capillary

177

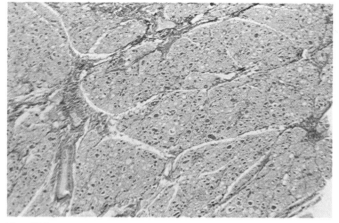

Fig. 408.

Fig. 408. Due to its lobular organization, together with the follicles closely resembling alveolar secretory units, at first sight the thyroid occasionally is confused with a lactating mammary gland (for differential diagnosis see Figs. 373–376 and Table 14). In most cases the follicles are found in various stages of activity ranging from totally emptied to completely filled with colloid (human thyroid). Mallory-azan staining. Magnification 38×.

Collapsed follicle

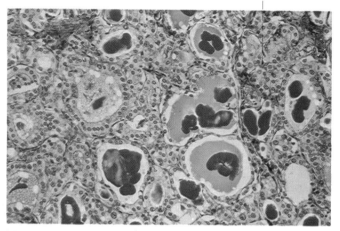

Fig. 409.

Fig. 409. At a higher magnification it becomes evident that the staining reaction of the colloid not only differs considerably in different follicles but also within the same. This is an expression of the varying water contents of the colloid that with increasing age is gradually reduced and progressively the colloid stains red with azo-carmine. Furthermore as most of the ordinary techniques used in preparing slides involve a total withdrawal of tissue water this leads to shrinkage of the colloid that thereby is retracted from the follicular epithelium to a varying extent (human thyroid). Mallory-azan staining. Magnification 150×.

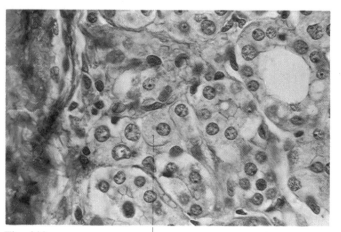

Fig. 410.

Parafollicular cell

Fig. 410. Several small and completely emptied follicles from a human thyroid among which can be seen a few parafollicular cells that produce the hormone thyrocalcitonin. These cells, now unanimously accepted as separate thyroidal elements, can easily be simulated by tangential sections through small follicles and hence are readily confused with them. Therefore their independent existence was long disputed. Mallory-azan staining. Magnification 480×.

Fig. 411. Complete midsagittal section through an isolated human parathyroid, whose identification does not offer any difficulties because in most cases it is sectioned with adhering thyroidal tissue. Only if an isolated specimen is sectioned might this be confused with the epiphysis cerebri. The differentiation is based (1) on the different sizes of the two organs (cf. Fig. 405), (2) on the considerably lesser amount of a more delicate interstitial connective tissue within the parathyroid, and (3) on the close attachment of clearly defined epitheloid parenchymal cells in the parathyroid. H. and phloxine staining. Magnification 38×.

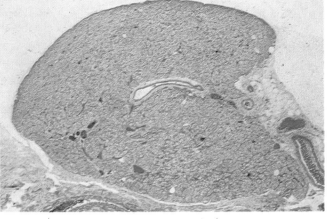

Fig. 411.

Fig. 412. The epithelioid nature of the parenchymal cells illustrated to a better advantage with a higher magnification (human parathyroid). The individual elements differ with regard to the intensity of their staining reactions. The two particularly acidophilic globules located at the left- and right-hand side of the lower third of the micrograph correspond to colloid droplets that are occasionally found also in this gland. Mallory-azan staining. Magnification 380×.

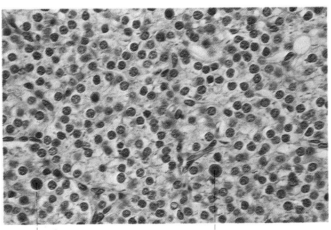

------ *Colloid droplet* ------   **Fig. 412.**

Light chief cell

Fig. 413. In the center of the micrograph can be seen one of the large oxyphil cells from a human parathyroid, whose cytoplasm shows only a weak acidophilic reaction and whose nucleus exhibits no signs of pyknosis as often found in these elements. They are a rare cell type and, in this specimen, surrounded by "light" and "dark" chief cells. The latter are assumed to represent the active secretory stage, while the former are particularly clearly outlined against each other because their cytoplasms remain nearly unstained. Mallory-azan staining. Magnification 960×.

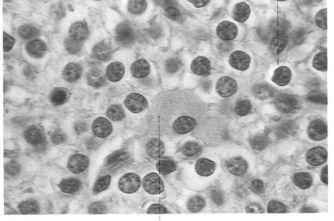

Fig. 413.   *Oxyphil cell*

179

Endocrine glands – The adrenal gland

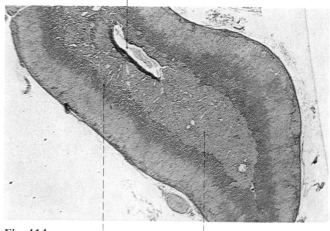

Medullary vein

Fig. 414. *Zona reticularis*   *Medulla*

Zona glomerulosa

Fig. 414. When viewed with the unaided eye or with the lowest-power objective, as illustrated here, the adrenals are characterized in cross sections by their organization into definite layers without these necessarily coinciding with the subdivision of the organ into a cortex and a medulla. An unbiased spectator would describe three layers in this specimen: (1) an outer faintly staining layer, (2) a middle darker stained zone and (3) an inner and paler area. Only the last corresponds to the adrenal medulla, while the other two represent the cortex (cf. also Fig. 415). A characteristic of the medulla is the large veins equipped with thick, highly muscular walls. Mallory-azan staining. Magnification 15×.

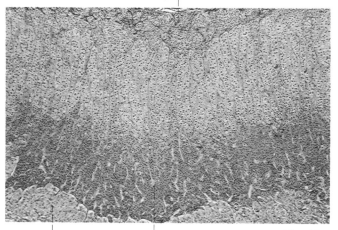

Medulla   *Zona reticularis*   **Fig. 415.**

Fig. 415. Only at a higher resolution, i.e., using a higher powered objective, can the three different zones of the adrenal cortex be identified. According to the arrangement of their cells they are known as (1) zona glomerulosa (cells being gathered into ovoid groups), (2) zona fasciculata (cells aligned in parallel cords) and (3) zona reticularis (cell cords forming a network). The zona reticularis stains particularly well and hence is often mistaken for the medulla (human adrenal). Mallory-azan staining. Magnification 48×.

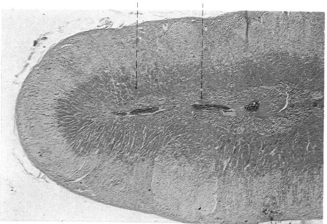

Zona reticularis   *Medulla*

Fig. 416.

Fig. 416. If a fresh adrenal is transferred into a fixative containing potassium bichromate, the medullary cells become brown and therefore are known as chromaffin cells (porcine adrenal). This reaction is due to the easy oxidation of epinephrine and norepinephrine contained within these cells in a granular form. Also in this specimen the zona reticularis stands out very clearly due to its darker stain. Nuclear fast red staining. Magnification 24×.

180

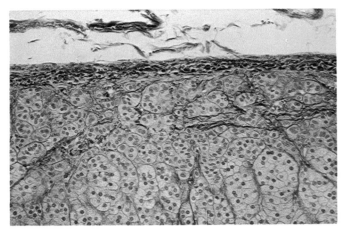

Fig. 417.

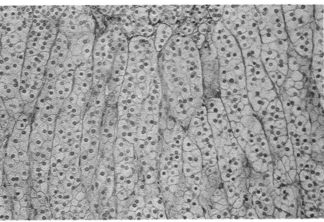

Fig. 418.

Details of the three cortical zones and the medulla from the same human adrenal.
All figures: Mallory-azan staining. Magnifications 150×.

Fig. 417. Subjacent to the delicate connective tissue capsule is found the narrow zona glomerulosa consisting of small ovoid groups of cuboidal cells. These cells are enveloped by a fine network of reticular fibers extending from the capsule throughout the adrenal cortex. The cells of the zona glomerulosa nearest the capsule are assumed to be only poorly differentiated elements forming the "cortical blastema."

Fig. 418. The cells of the zona fasciculata are arranged into cords running in parallel, and in ordinary preparations they give a vacuolated, spongy appearance (hence often called "spongiocytes", cf. also Fig. 42). This is due to the dissolving of their numerous lipid droplets by the usual technical procedures.

Fig. 419. The deeper staining cells of the zona reticularis are arranged into interconnecting cords forming a network (reticulum) containing numerous sinusoidal blood vessels. Note that along the lower margin of the micrograph the outermost layers of the medulla are seen.

Fig. 420. The medullary cells originate from the sympathetic primordia and hence correspond to a paraganglion. As usual in ordinary preparations, in this slide none of their intracytoplasmic granules can be seen. These are only prominent when oxidized, as with potassium bichromate, and thereby furnished with a brownish tinge (cf. Fig. 416). Due to this technique these cells are known as "chromaffin" or "pheochrome" cells (phaeos, Gr. = brown).

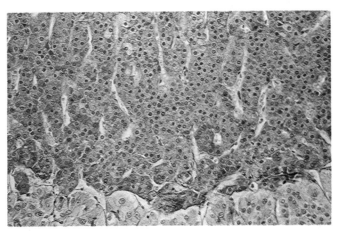

Fig. 419.

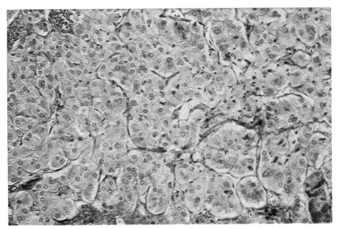

Fig. 420.

181

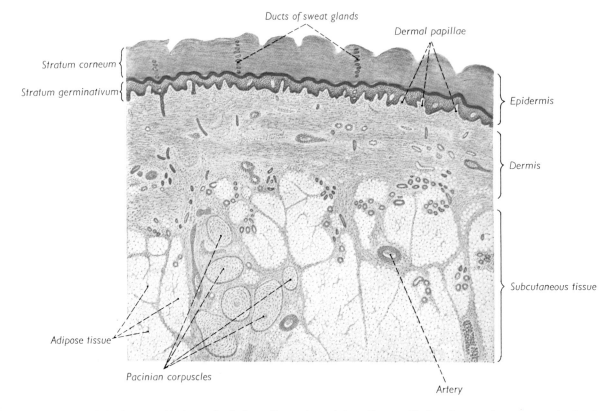

Fig. 421. The lamination of the human skin is particularly well seen in such heavily cornified regions as the palms and soles. It consists of two main parts: (1) the epidermis and (2) the underlying corium. The epidermis can be roughly subdivided into a superficial cornified layer (= stratum corneum) and a cellular layer (= stratum germinativum) beyond, with a deeper staining band (= stratum granulosum and stratum lucidum) between (skin from a human palm). The corium (= dermis) is composed of connective tissue, and its superficial or papillary layer serves as a mechanical device for a firm attachment of the epidermis. The deeper or reticular layer contains not only coarser collagenous fiber bundles, but the majority of the glands and blood vessels as well (camera lucida drawing). H.E. staining. Magnification 18 ×.

Duct of sweat gland

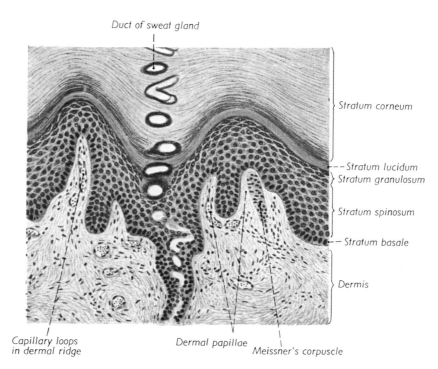

Stratum corneum

Stratum lucidum
Stratum granulosum

Stratum spinosum

Stratum basale

Dermis

Capillary loops
in dermal ridge

Dermal papillae

Meissner's corpuscle

Fig. 422. At a higher magnification the epidermis displays a more detailed lamination than described before: (1) the stratum basale consisting of columnar cells (hence also called str. cylindricum) is arranged in a single layer which together with the following (2) stratum spinosum forms the stratum germinativum. This is overlayed by the (3) stratum granulosum which, due to its deeply staining keratohyalin granules, stands out very clearly. Between this and the cornified superficial layer lies the highly refractile stratum lucidum. In the superficial stratum corneum the keratohyalin granules gradually merge with the tonofilaments, finally forming a filament-matrix complex, while the cell membranes become thickened and the nuclei together with the rest of the organelles vanish completely (camera lucida drawing). H.E. staining. Magnification 170 ×.

Arterial segment

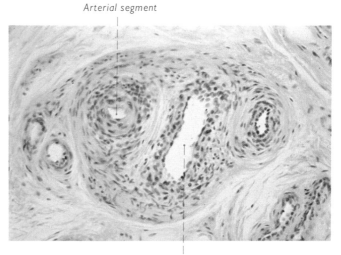

Fig. 423. Transverse section through an organ of Hoyer-Grosser in the subcutaneous tissue of a human finger tip. These special components of the microvascular bed consist of coiled arteriovenous anastomoses that posses a particularly rich innervation and are enclosed by a fibrous capsule like other organs. H.E. staining. Magnification 150 ×.

Venous segment

183

The integument – The epidermis

"Intercellular bridges" in the stratum spinosum

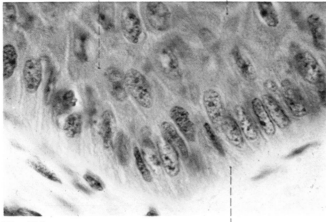

Fig. 424.

Delicate basal processes of stratum basale

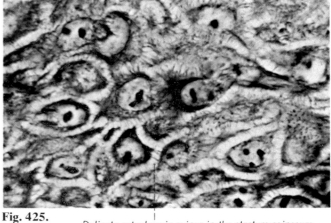

Fig. 425.

Delicate cytoplasmic spines in the stratum spinosum

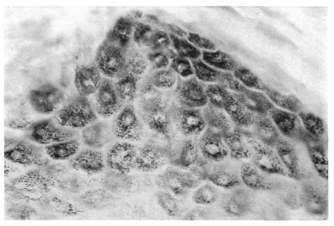

Fig. 426.

Epithelial nuclei

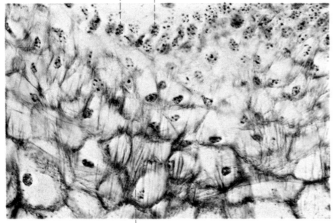

Fig. 427.

Tonofibrils

Fig. 424. The columnar (note nuclear shape) epithelial cells of the stratum basale extend with slender cytoplasmic processes into the underlying connective tissue to achieve a firmer attachment to it. Just above the basal cells the "cross bridges" between the prickle cells of the stratum spinosum are faintly visible (epidermis of a human finger tip). H.E. staining. Magnification 960×.

Fig. 425. Due to the shrinkage of the cells caused by the technical procedures and the numerous desmosomes by which the cells of the stratum spinosum are linked together these elements became studded with many spiny processes while being pulled apart. Hence their name "prickle" cells (condyloma accuminatum, man). Iron-hematoxylin staining. Magnification 960×.

Fig. 426. Tangential section through the stratum granulosum of the epidermis from the human finger tip. Note that the number of the keratohyalin granules, one of the precursors of the cornified substance, gradually increases toward the epithelial surface. H.E. staining. Magnification 380×.

Fig. 427. For the demonstration of the intracellular tonofibrils usually nonhuman material is used (epithelial matrix of the hoof from a bovine fetus). Each of these filamentous structures consists of finer subunits, the tonofilaments (cf. Fig. 37), and they serve as a cytoskeleton for each individual cell. Furthermore in their entirety they enhance the endurance of the epithelium as a whole for mechanical stresses due to being arranged along the lines of major mechanical stress. Iron-hematoxylin staining. Magnification 380×.

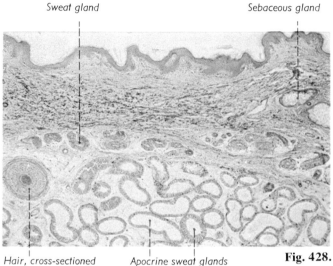

Sweat gland · Sebaceous gland

Hair, cross-sectioned · Apocrine sweat glands · **Fig. 428.**

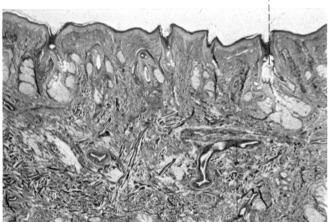

Tunica dartos

Fig. 429. · Apocrine sweat glands

Sebaceous gland

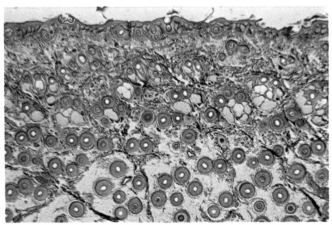

Fig. 430. · Skeletal muscle fibers · Artery

Defined areas of the integument such as the axillary region and the skin of the palms or soles, of the scalp, of the scrotum and the labia must be correctly identified as such because of their characteristic morphological features.

Fig. 428. The axillary skin shows a low and poorly cornified epithelium together with hairs and sebaceous and sweat glands. The most peculiar feature is the numerous and well-developed apocrine sweat glands. These are characterized by their wide lumina and the varying height of their secretory cells (for details cf. Figs. 437 and 438). Iron-hematoxylin and benzopurpurin staining (slightly faded). Magnification 38×.

Fig. 429. Skin of the human scrotum with the characteristic layer of smooth muscle (tunica dartos) that clearly distinguishes this cutaneous area from all others. In addition, apocrine and eccrine sweat glands together with sebaceous glands occur in varying numbers and occasional hairs may be found. Iron-hematoxylin staining. Magnification 38×.

Fig. 430. Skin from a human nostril. Typical for this area are the numerous sebaceous glands that are not associated with hairs. For detailed identifying characteristics see Table 11. Mallory-azan staining. Magnification 20×.

Fig. 431. The skin of the human scalp can easily be identified because of the large number of closely spaced hairs. As these are cut at different levels in a section parallel to the epithelial surface their cross sections will greatly vary in appearance (see also Fig. 432). Mallory-azan staining. Magnification 17×.

Fig. 431.

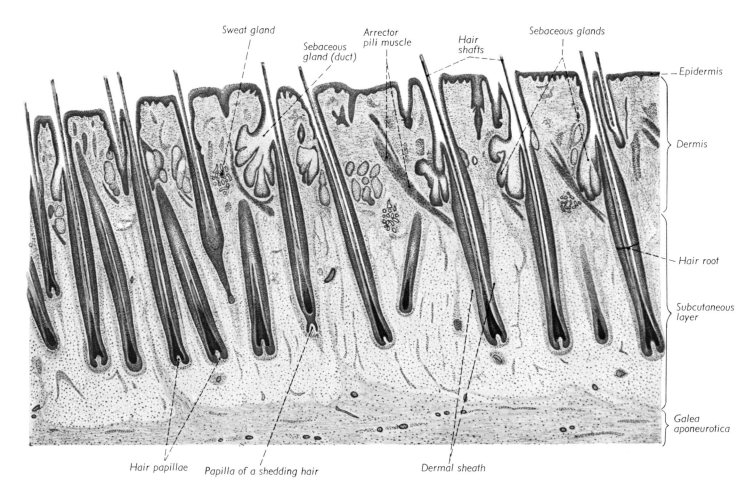

Sweat gland

Sebaceous
gland (duct)

Arrector
pili muscle

Hair
shafts

Sebaceous glands

Epidermis

Dermis

Hair root

Subcutaneous
layer

Galea
aponeurotica

Hair papillae   Papilla of a shedding hair

Dermal sheath

Fig. 432. Longitudinal sections through the hairs (human scalp) show their free ends, the shafts, projecting above the surface while their roots are embedded in deep, narrow invaginations consisting of an epithelial and a connective tissue sheath that together form the hair follicle (camera lucida drawing). H.E. staining. Magnification 40×.

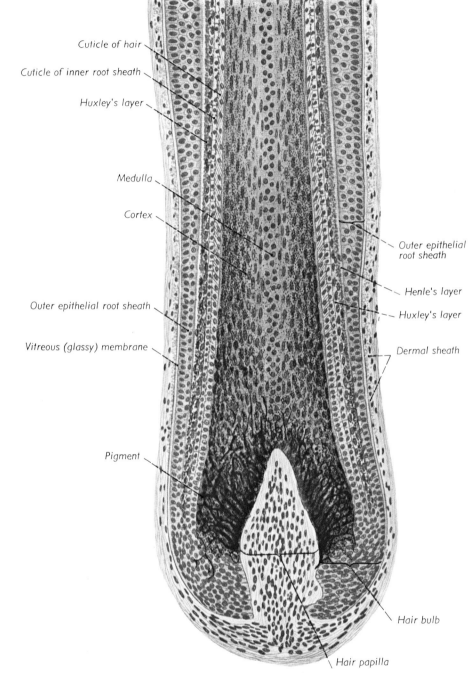

Cuticle of hair

Cuticle of inner root sheath

Huxley's layer

Medulla

Cortex

Outer epithelial root sheath

Vitreous (glassy) membrane

Pigment

Outer epithelial root sheath

Henle's layer

Huxley's layer

Dermal sheath

Hair bulb

Hair papilla

Fig. 433. At a higher magnification the epithelial part of the hair follicle displays its rather complex layering. It is subdivided into (1) an inner and (2) an outer epithelial root sheath. The inner layer consists of the cuticle of the root sheath which, by interdigitations with the hair cuticle, achieves a firm anchorage of the hair root within its sheath. This is followed by the Huxley's layer that consists of one or two rows of elongated cells onto which is attached the Henle's layer, which is formed by a single row of flattened cells. The outer epithelial root sheath is continuous with the stratum germinativum of the epidermis, and its outermost cylindrical cells are covered by the hyaline or vitreous membrane that is the inner layer of the connective tissue sheath of the hair follicle (camera lucida drawing). H.E. staining. Magnification 200×.

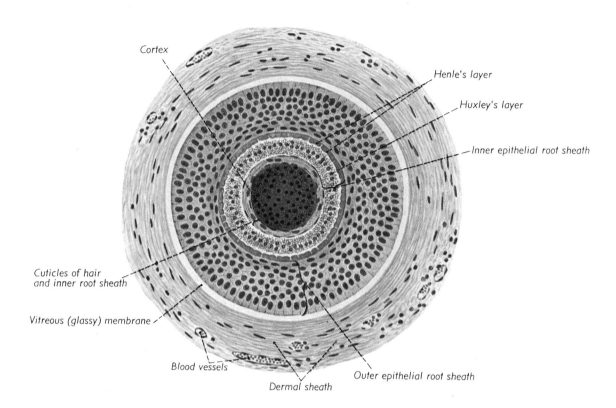

Cortex

Henle's layer

Huxley's layer

Inner epithelial root sheath

Cuticles of hair
and inner root sheath

Vitreous (glassy) membrane

Blood vessels

Dermal sheath

Outer epithelial root sheath

Fig. 434. Transverse section through the hair root showing its various sheaths (camera lucida drawing). Compare with Fig. 433. H.E. staining. Magnification 300×.

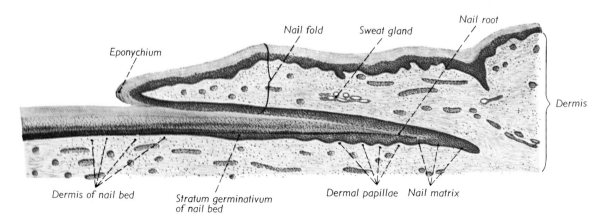

Eponychium

Nail fold

Sweat gland

Nail root

Dermis

Dermis of nail bed

Stratum germinativum
of nail bed

Dermal papillae

Nail matrix

Fig. 435. Longitudinal section of the proximal parts of a newborn's nail (camera lucida drawing). H.E. staining. Magnification 30×.

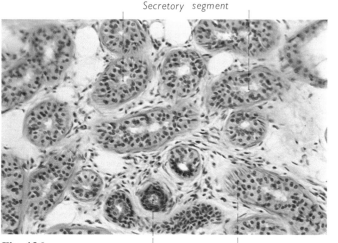

Secretory segment

Fig. 436.

Origin of excretory duct   Myoepithelial cells

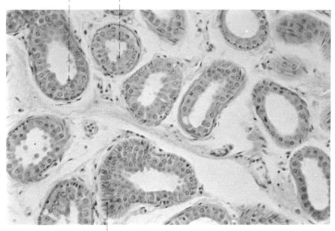

Cellular apices with secretion product

Myoepithelial cells   **Fig. 437.**

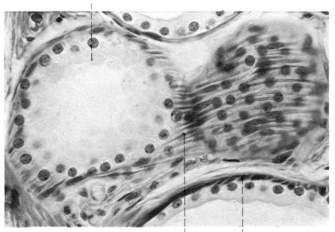

Cytoplasmic hoods formed by the apocrine secretion mechanism

Fig. 438.   Myoepithelial cells

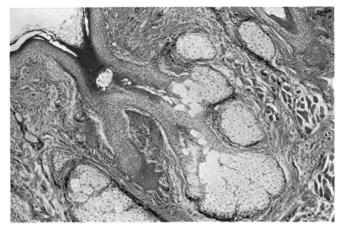

Fig. 439.

Fig. 436. The eccrine sweat glands are simple tubular glands whose distal parts are tightly coiled (cf. Fig. 82) and are predominantly located along the border line between dermis and the subcutaneous tissue. Their long excretory ducts possess a lumen that is less in diameter and is lined by deeper staining cells with closer spaced nuclei than found in the secretory portions (human finger tip). Please note myoepithelial cells that appear as discrete "stripes" at the sites of tangentially sectioned secretory tubules. H.E. staining. Magnification 150×.

Fig. 437. The apocrine sweat glands are branched alveolar glands that are only found in certain areas of the skin. They are characterized by the wide lumina in their secretory portions and by the varying height of their glandular epithelium. The latter has been assumed to represent the structural equivalent of the different steps in an apocrine secretion mechanism (human axillary skin). H.E. staining. Magnification 150×.

Fig. 438. In tangential sections of the secretory alveoli of the apocrine sweat glands the contractile, spindle-shaped myoepithelial cells can be demonstrated particularly clearly without being a unique feature of these glands (human ceruminous glands). Mallory-azan staining. Magnification 380×.

Fig. 439. The holocrine sebaceous glands also belong to the branched alveolar type, but their lumina are in most cases obstructed by masses of epithelial cells gradually transformed into the secretory product, the sebum. Mallory-azan staining. Magnification 60×.

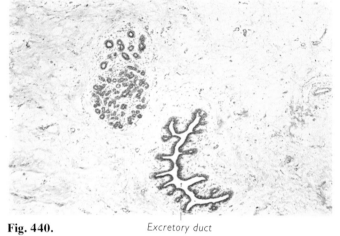

Fig. 440.
Excretory duct

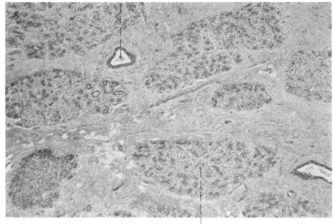

Excretory duct

Fig. 441.
Glandular lobule

Adipose tissue *Excretory duct*

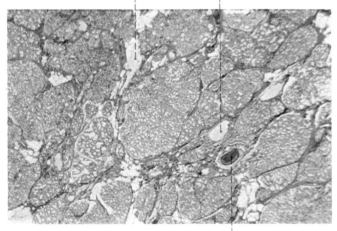

Fig. 442.
Artery stuffed with erythrocytes

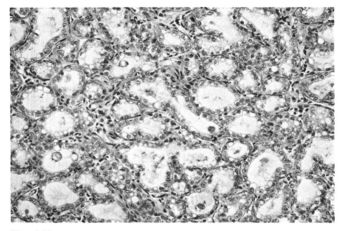

Fig. 443.

Fig. 440. Part of a resting mammary gland from a virgin consisting of a larger excretory duct and a cluster of secretory alveoli surrounded by a thin fibrous capsule. H.E. staining. Magnification 38×.

Fig. 441. The proliferating mammary gland of a pregnant woman. Under the influence of various hormones during pregnancy the epithelial tubules of the inactive gland begin to increase in number and size. Thereby they compress the remainder of the extremely reduced connective tissue into small strands that persist as the interlobular septa carrying the larger blood vessels and excretory ducts. H.E. staining. Magnification 38×.

Fig. 442. In the fully developed state the lactating mammary gland consists of 10–15 separate tubulo-alveolar glands, whose secretory portions vary considerably in size due to their different stages of activity. Profiles of the larger excretory ducts are regularly found within the interlobular connective tissue septa (for other identifying characteristics cf. Fig. 374 and Table 14). Mallory-azan staining. Magnification 34×.

Fig. 443. As the usual embedding procedures involve lipid solvents like alcohol, benzene, etc., the cells of the secretory active alveoli contain numerous vacuoles instead of fat droplets. Despite an occasional similarity of the alveolar contents with the thyroid colloid, the mammary alveoli can always be identified as such by their outlines being much more irregular than those of the thyroid follicles. Mallory-azan staining. Magnification 150×.

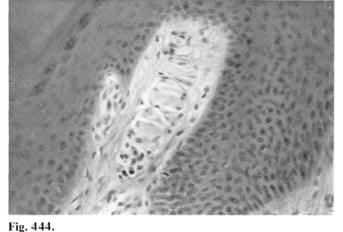

Fig. 444.

Fig. 444. Longitudinal section of a tactile corpuscle of Meissner from human finger tip. These are located in the dermal papillae, especially of the hairless skin. They consist of a stack of elongated, club-shaped connective tissue cells between which an afferent axon pursues its spiral course. H.E. staining. Magnification 240×.

Axon

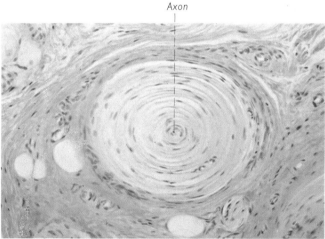

Fig. 445.

Fig. 445. Transverse section through a Pacinian corpuscle (human finger tip) that also serves as a receptor for mechanical stimuli. These are, however, predominantly found in the deeper layers of the subcutaneous connective tissue. They are composed of a centrally located single axon surrounded by a large number of concentric cellular lamellae separated from one another by interstices filled with a clear fluid. Iron-hematoxylin and benzopurpurin staining. Magnification 150×.

Connective tissue capsule

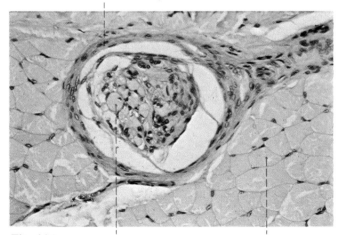

Fig. 446. Cross section of a muscle spindle from human m. lumbricalis. These receptors, like the two foregoing ones, possess a prominent connective tissue capsule that encloses a number of so-called intrafusal fibers. These are arranged parallel to the ordinary muscle fibers from which they differ by having a smaller diameter, a noncontractile midportion and a special innervation. Hematoxylin staining. Magnification 240 ×.

Fig. 446.   Intrafusal muscle fiber   Skeletal muscle fiber

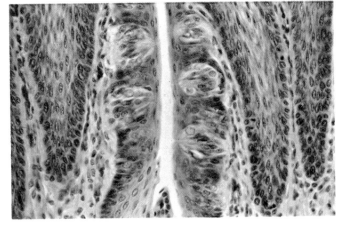

Fig. 447.

Fig. 447. Several taste buds located in the epithelium lining the trench between the foliate papillae (rabbit's tongue). Due to their poor stainability these sensory organs appear at low magnifications as cone-shaped translucencies within the darker staining epithelium. Iron-hematoxylin (Heidenhain) staining. Magnification 240×.

Nucleus of a neuroepithelial (taste) cell

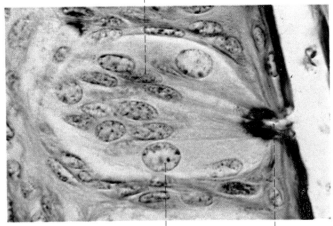

Fig. 448.

Nucleus of a supporting (sustentacular) cell   *Taste pore containing "taste hairs"*

Fig. 448. At higher magnifications two types of cells can be distinguished within the taste buds due to the different sizes of their nuclei (rabbit's foliate papilla). The one contains a large roundish nucleus and is called a sustentacular cell. Its apical portion does not regularly reach the taste pore. The other type is the taste cell that shows a more elongated and deeper staining nucleus and always reaches into the taste pore with an apical process, the "taste hair" (= bunch of slender microvilli). Due to the thickness of this section the latter appear as a homogeneous blackening along the bottom of the taste pore. Iron-hematoxylin (Heidenhain) staining. Magnification 960×.

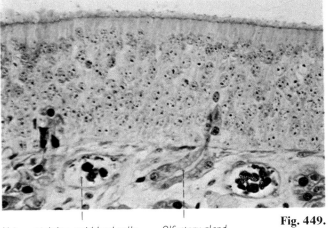

Vein containing red blood cells   *Olfactory gland*   **Fig. 449.**

Fig. 449. The pseudostratified columnar epithelium of the olfactory mucosa (canine regio olfactoria) can be distinguished from the respiratory epithelium because it is considerably thicker and contains no goblet cells. As human material is difficult to obtain in a well-preserved state, the olfactory mucosa of various animals is shown in many histologic courses instead. But like the human olfactory epithelium it contains (1) sustentacular and (2) olfactory cells that are of the nature of bipolar ganglion cells and are difficult to identify as such in routine preparations. Also in this specimen details of the apical processes of the olfactory cells such as the olfactory vesicles and kinocilia cannot be seen due to the mucous covering the epithelial surface and the thickness of the section. Iron-hematoxylin and benzo light bordeaux. Magnification 380×.

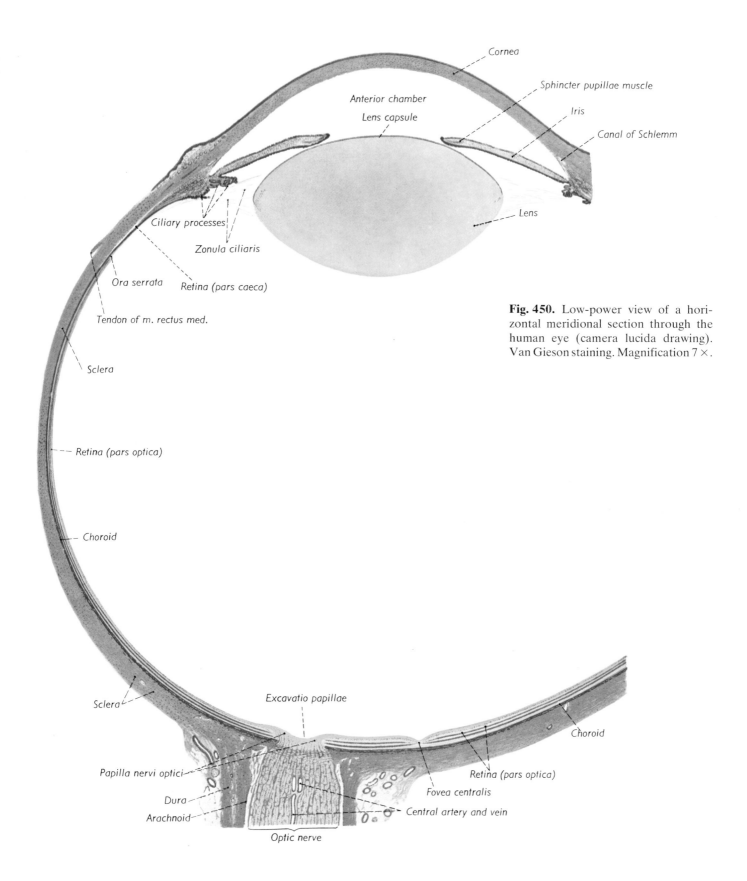

Cornea

Sphincter pupillae muscle

Iris

Canal of Schlemm

Anterior chamber

Lens capsule

Lens

Ciliary processes

Zonula ciliaris

Ora serrata

Retina (pars caeca)

Tendon of m. rectus med.

Sclera

Retina (pars optica)

Choroid

Fig. 450. Low-power view of a horizontal meridional section through the human eye (camera lucida drawing). Van Gieson staining. Magnification 7 ×.

Sclera

Excavatio papillae

Choroid

Retina (pars optica)

Papilla nervi optici

Dura

Arachnoid

Fovea centralis

Central artery and vein

Optic nerve

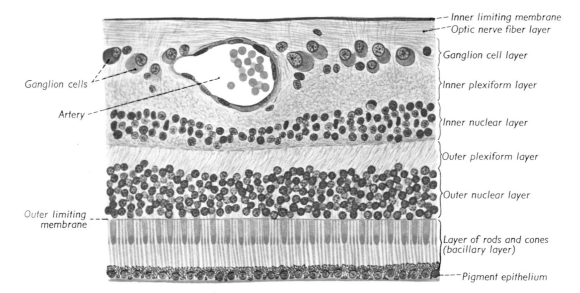

Ganglion cells
Artery
Outer limiting membrane

Inner limiting membrane
Optic nerve fiber layer
Ganglion cell layer
Inner plexiform layer
Inner nuclear layer
Outer plexiform layer
Outer nuclear layer
Layer of rods and cones (bacillary layer)
Pigment epithelium

Fig. 451. The light-sensitive part of the retina (= pars optica retinae) displays a complex stratification that can best be understood as a sequence of three different interconnected neurons. Considering them in the order of conduction we find that the outermost layer is that of the first neuron, namely the photoreceptors (= rod and cone cells). Inwardly there follow two layers of nerve cells that together with their cytoplasmic processes represent the second and third neuron of the optic tract. As both the nuclei of these three neurons and their processes are located at well-defined levels within the retina, the latter appears as "stratified." The two "nuclear" and the "ganglion cell" layers contain the cell bodies and nuclei (1) of the rod and cone cells (= outer nuclear layer), (2) of the bipolar neurons (= inner nuclear layer) and (3) of the multipolar ganglion cells of the optic nerve. Both "plexiform" layers are composed of the processes of the adjoining nerve cell layers in such a fashion that in the outer plexiform layer the axons of the rod and cone cells (first neuron) make synaptical contacts with the dendrites of the bipolar nerve cells (second neuron), while in the inner plexiform layer the axons of the latter (second neuron) form axodendritical synapses with the ganglion cells (third neuron) of the optic nerve. The innermost retinal layer contains the axons that converge toward the papilla, thus finally forming the n. opticus. The inner limiting membrane is formed by the apposition of the expanded ends of the slender processes of specific glial cells (= supporting Müller cells), where as the outer "limiting membrane" corresponds to an elaborate system of intercellular adhesive devices established between the Müller cells and the outer segments of the photoreceptors. In routine preparations Müller cells cannot be identified (camera lucida drawing). H.E. staining. Magnification 400×.

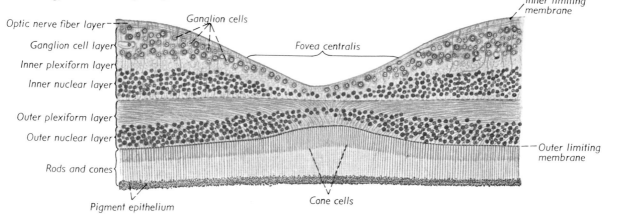

Optic nerve fiber layer
Ganglion cell layer
Inner plexiform layer
Inner nuclear layer
Outer plexiform layer
Outer nuclear layer
Rods and cones
Pigment epithelium

Ganglion cells
Fovea centralis
Inner limiting membrane
Outer limiting membrane
Cone cells

Fig. 452. Section through center of the fovea centralis within the macula lutea (= region of most acute vision), where the inner layers of the retina deviate, thus allowing the light to reach more directly the photoreceptors, which in this area are represented exclusively by cones (camera lucida drawing). H.E. staining. Magnification 175×.

194

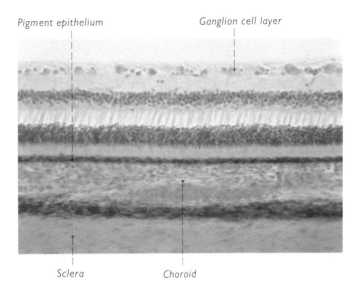

Pigment epithelium   Ganglion cell layer

Sclera   Choroid

Fig. 453. Original micrograph of a human retina in situ with the adjoining pigment epithelium, the choroid and the inner parts of the sclera. For the nomenclature of the various layers see Figs. 450 and 451. H.E. staining. Magnification 240×.

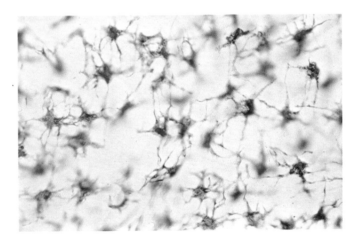

Fig. 454. Tangential section through a human cornea to illustrate the highly branched fibroblasts (keratocytes) interspersed in the substantia propria by impregnating the cornea as a whole with gold chloride. Magnification 240× (specimen courtesy of Prof. H. J. Clemens).

Organs of the special senses – The eye

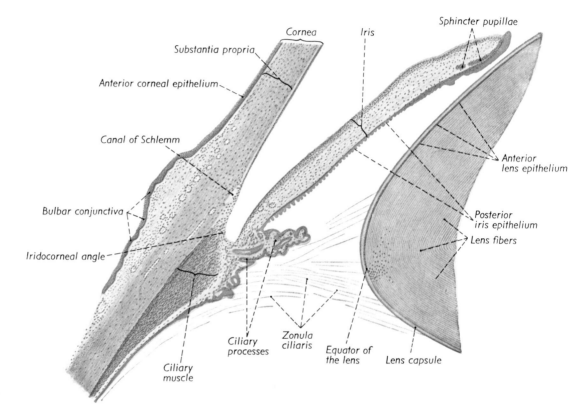

Fig. 455. Left half of a horizontal meridional section through the anterior part of the eyeball (cf. Fig. 450) showing the ciliary body together with the anterior and posterior chamber, the iris, lens and corneal rim (camera lucida drawing). H.E. staining. Magnification 35×.

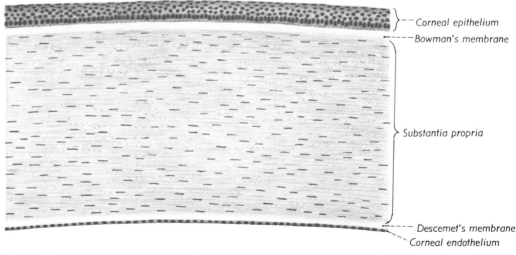

Fig. 456. The cornea normally contains no blood vessels and its stroma (= substantia propria) consists mainly of connective tissue fibers with modified fibroblasts interspersed, of which only the nuclei can be seen. Their branching cell bodies can be recognized in special preparation (e.g., by impregnation with gold see Fig. 454). The isolated cornea is a widely used specimen for the simultaneous demonstration of a stratified and a simple squamous epithelium (camera lucida drawing). H.E. staining. Magnification 80×.

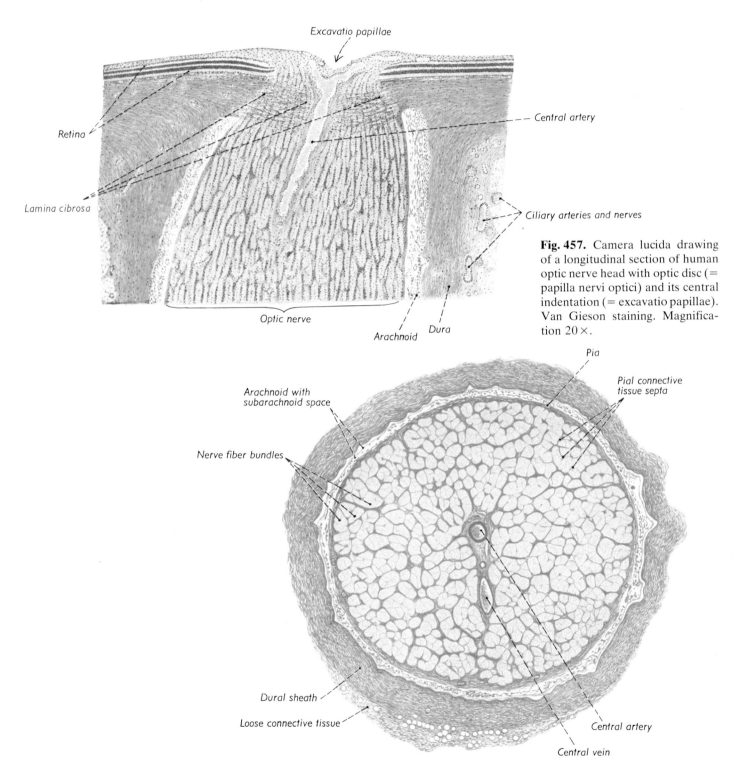

Excavatio papillae

Retina

Lamina cibrosa

Central artery

Ciliary arteries and nerves

Fig. 457. Camera lucida drawing of a longitudinal section of human optic nerve head with optic disc (= papilla nervi optici) and its central indentation (= excavatio papillae). Van Gieson staining. Magnification 20×.

Optic nerve

Arachnoid

Dura

Pia

Pial connective tissue septa

Arachnoid with subarachnoid space

Nerve fiber bundles

Dural sheath

Loose connective tissue

Central artery

Central vein

Fig. 458. Cross section of the optic nerve, which as a part of the brain is ensheathed by the three meninges, including a subarachnoid space. As the central artery and vein of the retina enter the nerve only at a distance of about a half inch behind the eye, these are never found in sections cut proximal from their entry and hence should not be considered as the only and most vital morphological criterion for the identification of the optic nerve (camera lucida drawing). Van Gieson staining. Magnification 22×.

197

Organs of the special senses – The eye

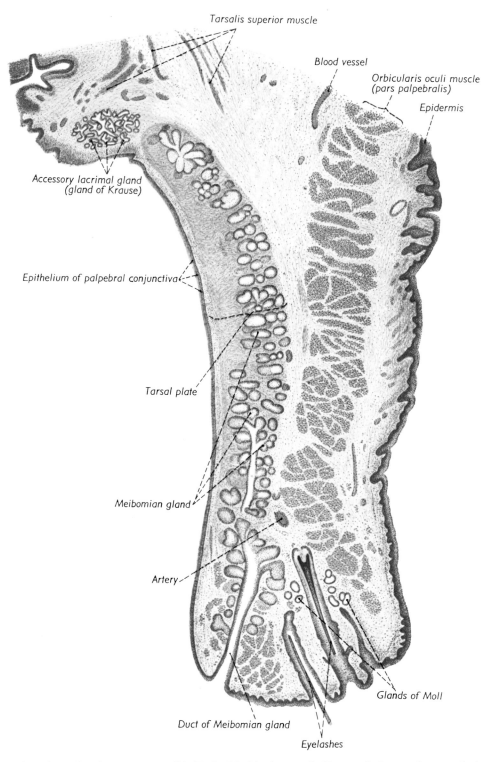

Tarsalis superior muscle

Blood vessel

Orbicularis oculi muscle
(pars palpebralis)

Epidermis

Accessory lacrimal gland
(gland of Krause)

Epithelium of palpebral conjunctiva

Tarsal plate

Meibomian gland

Artery

Glands of Moll

Duct of Meibomian gland

Eyelashes

Fig. 459. Vertical section through a human upper lid. Embedded in its tough fibrous skeleton, the tarsal plate, are sebaceous (= Meibomian) glands arranged in a single row with their long axis perpendicular to the lid margin. Contrary to these the apocrine sweat glands (Moll) are found close to the eyelashes. The upper border of the tarsus serves for the attachment of the involuntary superior tarsal muscle of Müller, whose tone keeps the lids open (camera lucida drawing). For further identifying characteristics see Table 11. H.E. staining. Magnification 17×.

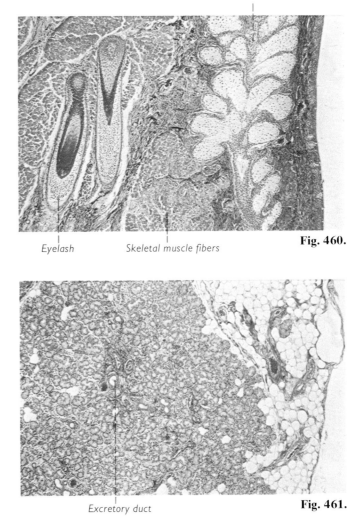

Meibomian gland

Fig. 460. Detail of human eyelid (vertical section) near the lid margin. At the right is seen the low palpebral conjunctival epithelium followed by a profile of a Meibomian gland, the cross-sectioned fibers of the orbicularis oculi muscle and parts of two eyelashes. Mallory-azan staining. Magnification 38×.

Eyelash   Skeletal muscle fibers   **Fig. 460.**

Fig. 461. Contrary to all the other serous glands, e.g., parotid and pancreas, the human lacrimal gland even at low magnifications displays the lumina of its secretory portions. According to their shape it has to be classified as a tubulo-alveolar gland, which furthermore is characterized by the lack of an elaborate duct system; hence only intra- and interlobular excretory ducts are found. For further identifying characteristics cf. Fig. 266 and Table 12. Mallory-azan staining. Magnification 38×.

Excretory duct   **Fig. 461.**

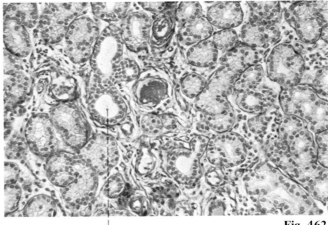

Fig. 462. The secretory cells of the alveoli regularly show spherical nuclei similar to those of serous alveoli of the parotid gland. The connective tissue interstices are richly cellular with numerous lymphocytes and small groups of plasma cells (cf. Fig. 101). Mallory-azan staining. Magnification 150×.

Excretory duct   **Fig. 462.**

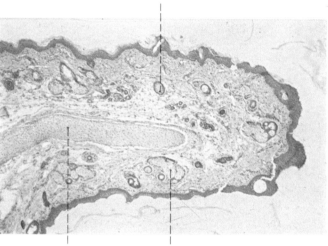

Undifferentiated primitive hair

Elastic cartilage   *Sebaceous gland*

Fig. 463. Horizontal section through the auricle of a human newborn. It consists of a plate of elastic cartilage covered on all sides by a thin skin with hair primordia and sebaceous glands. Borax carmine staining. Magnification 38 ×.

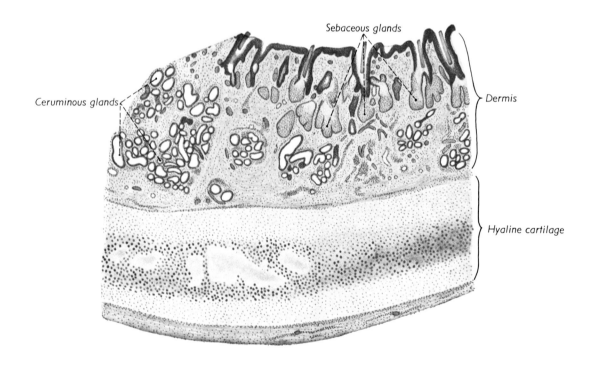

Sebaceous glands

Ceruminous glands

Dermis

Hyaline cartilage

Fig. 464. Part of a cross section through the cartilaginous portion of a human external auditory meatus (camera lucida drawing). It is lined by skin which not only shows hairs associated with sebaceous glands, but numerous profiles of large alveolar apocrine glands, the ceruminous glands, that are peculiar to this portion of the skin (for details cf. Figs. 87 and 438). H.E. staining. Magnification 16 ×.

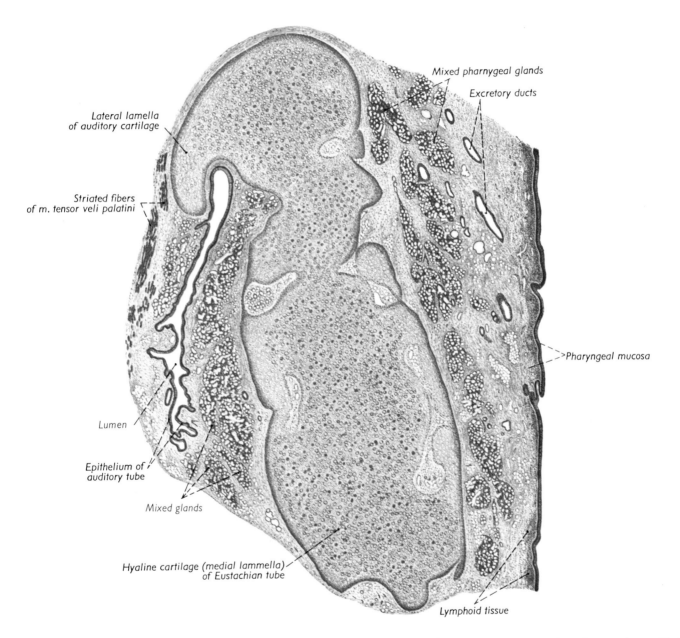

Mixed pharnygeal glands

Excretory ducts

Lateral lamella of auditory cartilage

Striated fibers of m. tensor veli palatini

Pharyngeal mucosa

Lumen

Epithelium of auditory tube

Mixed glands

Hyaline cartilage (medial lammella) of Eustachian tube

Lymphoid tissue

Fig. 465. Cross section through the cartilaginous part of the auditory (Eustachian) tube. Its mucosa consists of a pseudostratified columnar ciliated epithelium with goblet cells interspersed and a lamina propria showing an increasing number of aggregates of lymphatic tissue while approaching the inner orifice. The mucosal glands are seromucous in nature, and the cartilage is predominantly elastic (camera lucida drawing). H.E. staining. Magnification 13×.

Spiral cochlear ganglion

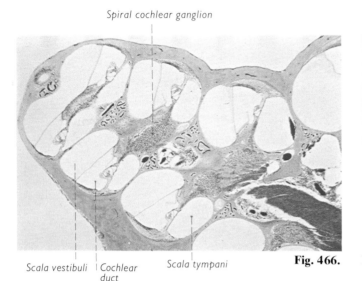

Scala vestibuli | Cochlear duct | Scala tympani | **Fig. 466.**

Stria vascularis   Vestibular membrane

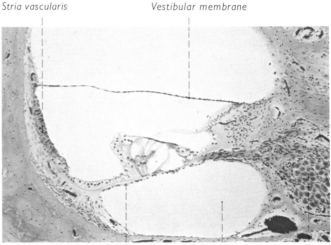

Fig. 467.   Basilar membrane   Scala tympani   Spiral cochlear ganglion

Outer tunnel   Outer hair cells   Tectorial membrane

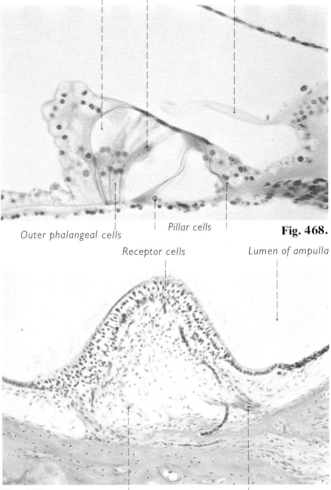

Outer phalangeal cells   Pillar cells   **Fig. 468.**

Receptor cells   Lumen of ampulla

Fig. 469.   Myelinated nerve fibers

Fig. 466. Axial section through the osseous cochlea of a guinea pig that in man makes about two and one-half turns around a conical axis, the modiolus. The latter contains the auditory nerve together with the regularly spaced profiles of the cross-sectioned spiral ganglion (consisting of bipolar ganglion cells) at both sides. At their level a bony shelf projects from the modiolus, the osseous spiral lamina, and radiates towards the membranous labyrinth. Mallory-azan staining. Magnification 24×.

Fig. 467. Cross section through one turn of the osseous canal of a guinea pig's cochlea shows three fluid-filled cavities of which the centrally located one corresponds to the cochlear extension of the membranous labyrinth. This cochlear duct contains the endolymph and is accompanied by two perilymphatic spaces running parallel, a lower scala tympani and an upper scala vestibuli. Against the latter the thin vestibular (Reissner's) membrane serves as the upper wall of the cochlear duct, while its lower wall mainly consists of the membranous spiral lamina. The outer wall is formed by the stria vascularis, which is richly supplied with capillaries and thought to produce the endolymph. Mallory-azan staining. Magnification 96×.

Fig. 468. Higher magnification of the auditory receptor, the organ of Corti, which consists of sensory (hair) cells and various supporting cells (pillar, phalangeal and border cells, cells of Hensen and of Claudius). Clearly visible in this micrograph are the three cavities extending the length of the cochlea: the inner tunnel, the space of Nuel and the outer tunnel (from right to left). In addition the following types of cells can be identified: inner and outer pillars (more eosinophilic), the outer phalangeal and outer hair cells and the supportive cells of Hensen and of Claudius. The epithelium covering the spiral limbus is continuous with both the tectorial membrane and the epithelial lining of the internal spiral sulcus. Mallory-azan staining. Magnification 240×.

202

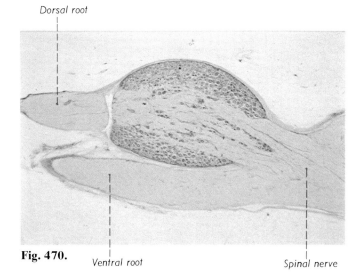

Dorsal root

Fig. 470. Longitudinal section through a canine spinal ganglion. Aggregates of sensory neurons are seen within the dorsal root (at the left side of the micrograph) shortly before joining the anterior root to form the spinal nerve (seen at the right side). Centrally the ganglion is bisected by myelinated nerve bundles running longitudinally. Cresyl violet staining. Magnification 21×.

Fig. 470. Ventral root   Spinal nerve

Fig. 471. The sensory neurons are mainly situated at the periphery of the ganglia, and most of them are of the "unipolar" variety whose central process forms the dorsal or afferent root of the spinal nerve. Among these rather large and more or less spherical cells darker staining elements can be seen that are richer in lipids and are believed to serve for the conduction of the protopathic sensibility (canine spinal ganglion). Cresyl violet staining. Magnification 120×.

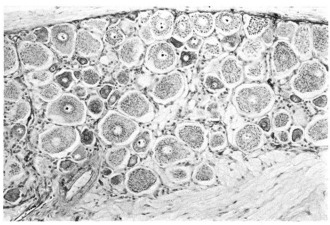

Fig. 471.

Fig. 472. Each of the unipolar neurons is invested by flattened peripheral glial cells that are akin to the Schwann cells, and quite often these "satellite cells" are separated from the neuronal soma by an artificial cleft (shrinkage). The Nissl substance of these ganglion cells is in the form of homogeneously distributed fine granules rather than in the shape of coarser chromophilic clumps. Note the numerous axon hillocks. Cresyl violet staining. Magnification 150×.

◄

Fig. 469. Crista ampullaris of a semicircular canal from the same specimen as shown in the preceding micrograph. The cupula has been lost in this section, but the myelinated fibers of the vestibular nerve are clearly visible in the connective tissue core of the crista. Mallory-azan staining. Magnification 150× (specimen for Figs. 466–469 courtesy of Dr. L. Thorn).

Axon hillock

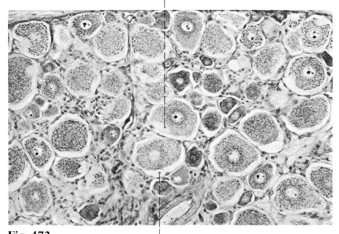

Fig. 472. Cleft between ganglion and satellite cells caused by shrinkage

203

Small group of ganglion cells

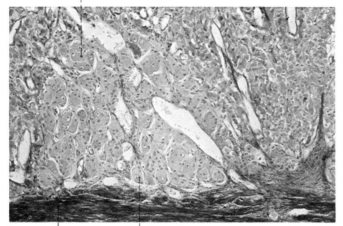

Smooth muscle from media
of thick walled medullary vein   Ganglion cell   **Fig. 473.**

Fig. 473. Autonomic ganglion from human adrenal medulla. These microscopically small aggregates of multipolar autonomic neurons are particularly frequent in this region because the medulla as a derivative of the sympathetic primordium belongs to the group of chromaffin paraganglia. The neurons can easily be identified by the large size of their cell bodies and nuclei regularly showing a prominent nucleolus. Mallory-azan staining. Magnification 95 ×.

Bundle of nerve fibers of the myenteric plexus   Smooth muscle

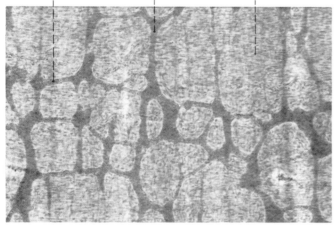

Fig. 474.

Fig. 474. Flat preparation of the myenteric (Auerbach's) plexus situated between the inner and outer muscular layer of the intestine. At the intersections of this network that is formed by unmyelinated nerve bundles of different size, small groups of autonomic (parasympathetic) ganglion cells can be found. Supravital staining with methylene blue. Magnification 95 ×.

Bundle of autonomic nerve fibers
with two small ganglion cells   Ganglion cell

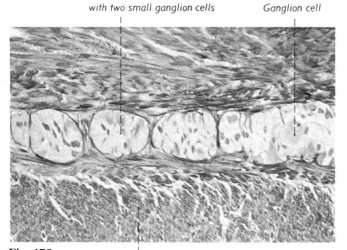

Fig. 475.   Smooth muscle, cross-sectioned

Fig. 475. Cross section of the myenteric plexus (Auerbach) exhibiting a few small ganglion cells located in between the unmyelinated autonomic nerves (human colon). Mallory-azan staining. Magnification 240 ×.

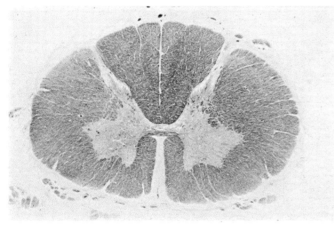

Fig. 476.

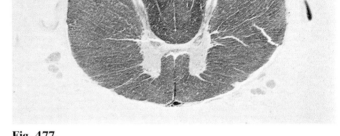

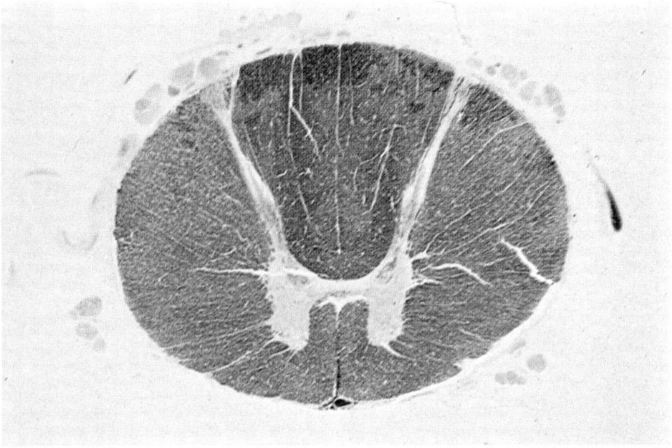

Fig. 477.

Figs. 476–478. Transverse sections through a cervical, thoracic and a lumbal segment of the human spinal cord that are all stained with the same technique for myelin and are all shown under identical magnifications (6×). By blackening the myelinated nerve fibers in the white matter this appears darker than the gray matter with its low amount of fibers. Note the diversities in size and shape of the gray matter and its lateral horn in the thoracic segment (Fig. 477). For nomenclature see Fig. 480.

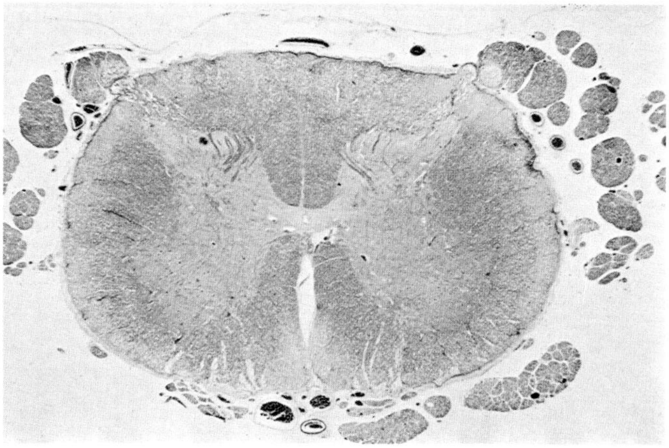

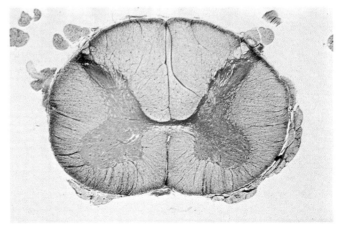

Fig. 478.

Fig. 479. Cross section through a cervical segment of a human spinal cord treated with a silver impregnation technique to illustrate the neurofibrils. If stained with the Nissl method that exclusively colors the nervous cells, the entire specimen appears almost colorless when viewed with the naked eye. To improve orientation in these cases search with the lowest-power objective for one of the anterior horns that are particularly rich in cells. Silver impregnation after Schultze-Stöhr. Magnification 6×.

Fig. 479.

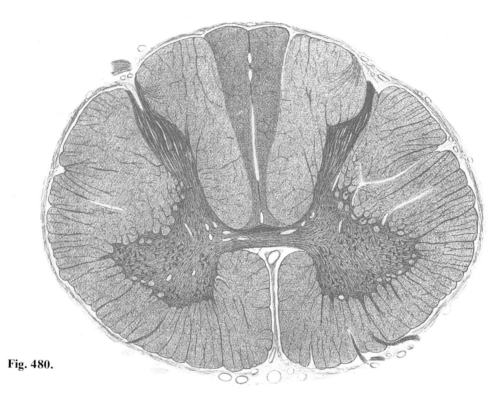

Fig. 480.

Figs. 480 and **481.** Cross sections through the cervical and lumbal intumescence of a human spinal cord. The white matter is subdivided into (1) the dorsal funiculus (dorsal or posterior white column) lying between the posterior horn and the dorsal median septum, (2) the lateral funiculus (lateral white column) lying between the anterior and posterior horns and roots and (3) the ventral funiculus (ventral or anterior white column) located between the anterior horn and the ventral median fissure of the spinal cord. Carmine staining (Fig. 480) and carmine staining combined with Weigert's stain for myelin. Magnification 8 × and 11 ×.

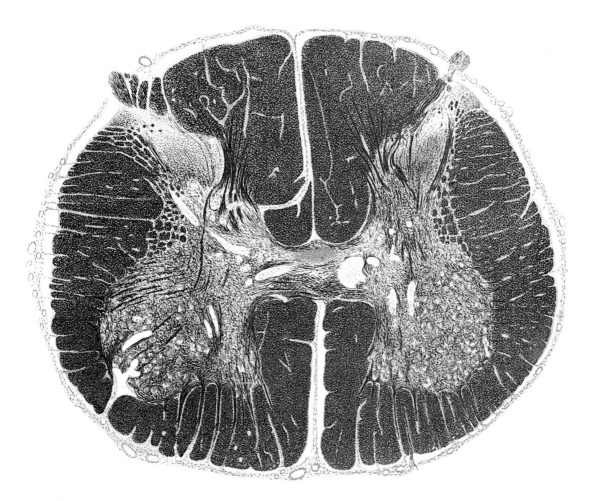

Fig. 481.

White matter

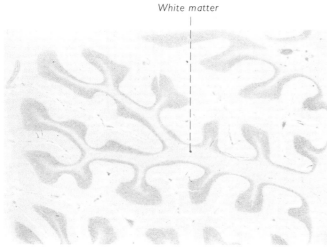

Fig. 482.

White matter

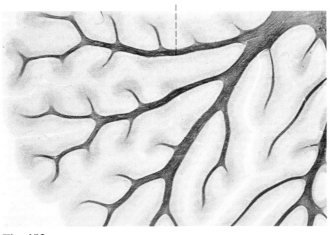

Fig. 483.

White matter

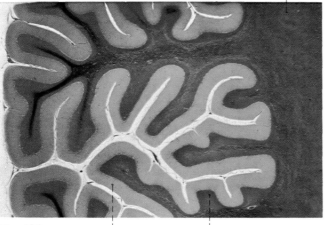

Fig. 484.   Molecular layer   Granular layer

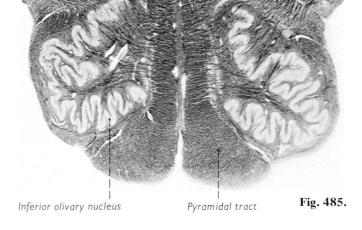

Inferior olivary nucleus   Pyramidal tract   **Fig. 485.**

Fig. 485. Transverse section through a human medulla approximately at the level of the upper third of the olive. Due to the stain for myelinated nerves, the intensely folded band of the olivary nucleus appears almost colorless while the cross-sectioned pyramidal tracts are distinctly outlined as massive fiber bundles positioned close to the midline. Staining: Weigert's method for myelin. Magnification 6×.

Figs. 482–484. Sagittal sections through the cortex of the human cerebellar vermis shown with the same magnifications (7×) but with different stains.

The highly cellular granular layer stands out particularly clearly as a grayish-blue band with the Nissl method that stains exclusively the nervous cells together with the glial nuclei, while a stain for myelin (Fig. 483) distinctly outlines the centrally positioned white matter and its finer ramifications. Only a combination of a stain for cells with a stain for myelin (Fig. 484) allows for both a clear illustration of the layering of the cerebellar cortex together with its demarcation from the central white matter. The cortex is approximately 1 mm thick and consists of (1) an outer molecular layer rather poor in cells (stained a yellow orange), (2) an intermediate layer of the Purkinje cells and (3) an inner granular layer (stained a reddish brown) that is rich in cells and borders upon the white matter. Stainings (from top to bottom): Nissl method, Weigert method for myelin, carmine combined with the Weigert method. Magnification 7×.

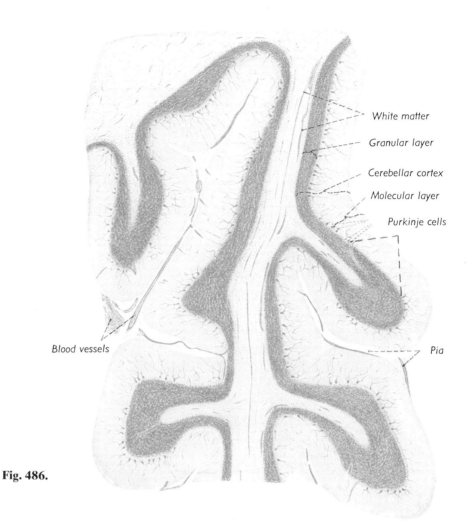

White matter

Granular layer

Cerebellar cortex

Molecular layer

Purkinje cells

Blood vessels

Pia

Fig. 486.

Fig. 486. Low-power view to demonstrate the lamination of the cerebellar cortex with a simple cell stain (camera lucida drawing). Carmine staining. Magnification 20×.

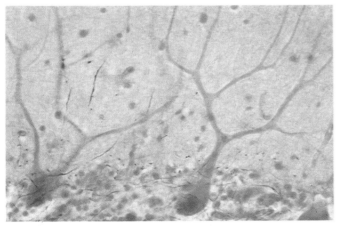

Fig. 487.

Fig. 487. A higher magnification of Fig. 485 better illustrates the thick, fan-shaped dendrites of the Purkinje cells that reach up to the cerebellar surface. The corresponding axon originates at the lower pole of the perikaryon and then traverses the granular layer to terminate upon one of the cerebellar nuclei. Staining: Weigert's method and carmine. Magnification 240×.

209

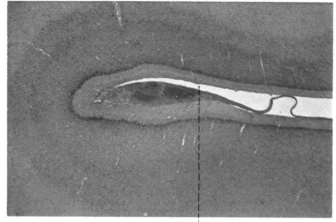

Fig. 488.

Central sulcus

Fig. 488. Most of the human cerebral cortex is six-layered and is known as the "isocortex" (neopallium). Even the brain of a human fetus displays this kind of lamination and its variations as well. The precentral gyrus (motor area) is seen in the upper half of the micrograph with the postcentral gyrus in the lower half. H.E. staining. Magnification 22×.

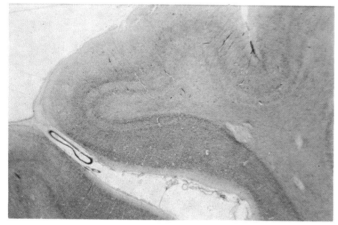

Fig. 489.

Fig. 489. Prominent lamination of the isocortex as seen in the human motor cortex (= gyrus precentralis). Due to the prevalence of the two pyramidal layers (the deeper staining cellular bands) and the poorly developed granular layers, this area it is known as the "agranular" type of the isocortex. Cresyl violet staining. Magnification 10×.

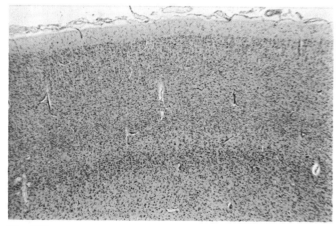

Fig. 490.

Fig. 490. At a higher magnification it becomes evident that the individual layers merge with each other. Together with the superficial and always faintly staining molecular layer (= lamina I) the inner pyramidal layer (= lamina V) stands out as a lighter staining band sandwiched between two highly cellular layers, the "outer pyramidal" (= lamina III) above and the "multiform" layer (= lamina VI) below. Cresyl violet staining. Magnification 33×.

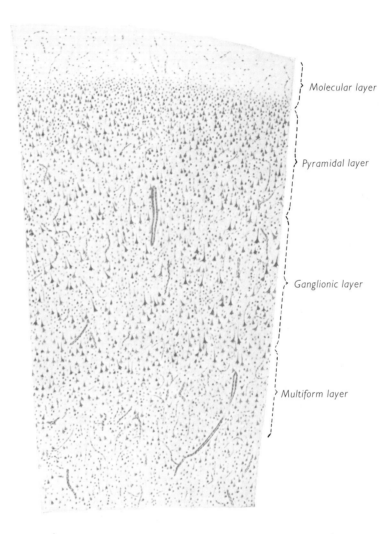

Molecular layer

Pyramidal layer

Ganglionic layer

Multiform layer

Fig. 491. Slightly schematic drawing of the cellular layers of the human motor cortex in which the inner granular layer is nearly lacking. Hence the two pyramidal layers govern the specimens of this cortical area, which therefore belongs to the "agranular" type. Carmine staining. Magnification 20 ×.

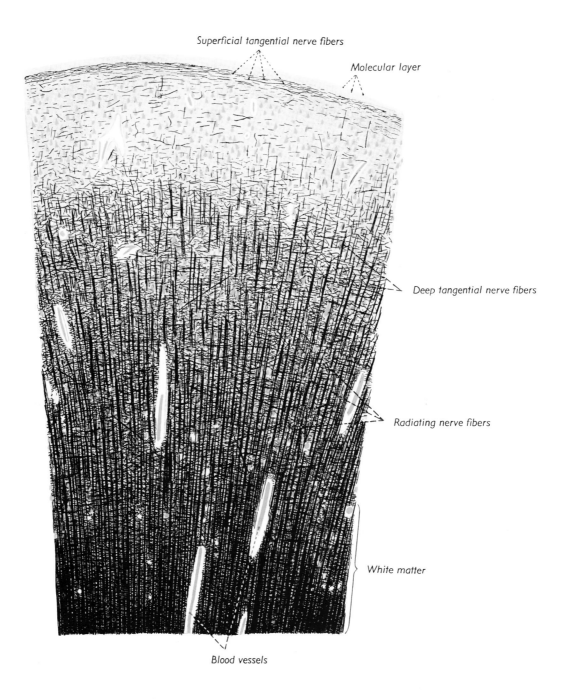

Superficial tangential nerve fibers

Molecular layer

Deep tangential nerve fibers

Radiating nerve fibers

White matter

Blood vessels

Fig. 492. When stained for the numerous myelinated nerve fibers existing in the cerebral cortex it becomes evident that the cortex not only possesses a distinct lamination with regard to its cellular components, but that, furthermore, the arrangement and distribution of the nerve fibers result in a structural feature known as the myeloarchitecture. In this the radiating fibers represent the bundles that ascend and descend to and from the cortex (camera lucida drawing). Staining: Weigert's method for myelin. Magnification 50 ×.

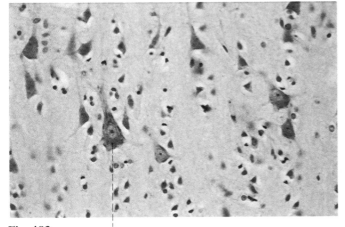

Fig. 493. *Giant pyramidal cell*

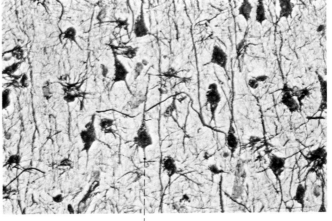

Fig. 494. *Axon of a pyramidal cell*

Fig. 493. Giant pyramidal cell of Betz from lamina V of the human motor cortex. Cresyl violet staining. Magnification 240×.

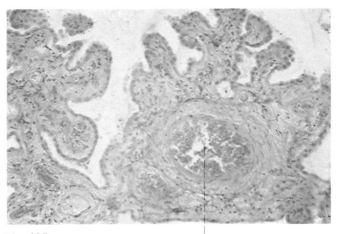

Fig. 495. *Vein with erythrocytes*

Fig. 494. Pyramidal cells and astrocytes together with their processes shown in a silver impregnated human cerebral cortex. The axon of the pyramidal cells is that process that originates at the midpoint of the cell base and courses straight downward. Silver impregnation. Magnification 300× (specimen courtesy of Prof. H. J. Clemens).

Veins filled with erythrocytes

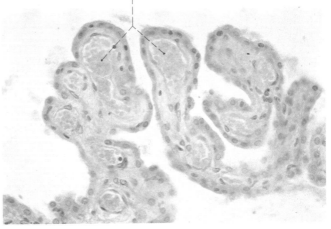

Figs. 495 and 496. Parts of a choroid plexus from the human lateral ventricle. These highly vascular and intensely folded connective tissue lamellae project freely into the ventricles, and they produce together with their covering epithelium (simple, cuboidal) the cerebrospinal fluid (CSF). In addition, the epithelium functions as a selective barrier for materials transported from the blood into the CSF and therefore it is part of the blood-CSF barrier system. At higher magnifications (Fig. 496) individual villi of the choroid plexus can easily be confused with placental villi. The structures, however, can be distinguished from one another by the following criteria: (1) the much looser fibrous stroma within the placental villi, (2) the bilayered epithelium in the early placenta that later becomes greatly attenuated and (3) the regularly occurring fibrinoid within the intervillous space (cf. Figs. 391–393). H.E. staining. Magnifications 120× and 240×.

Fig. 496.

213

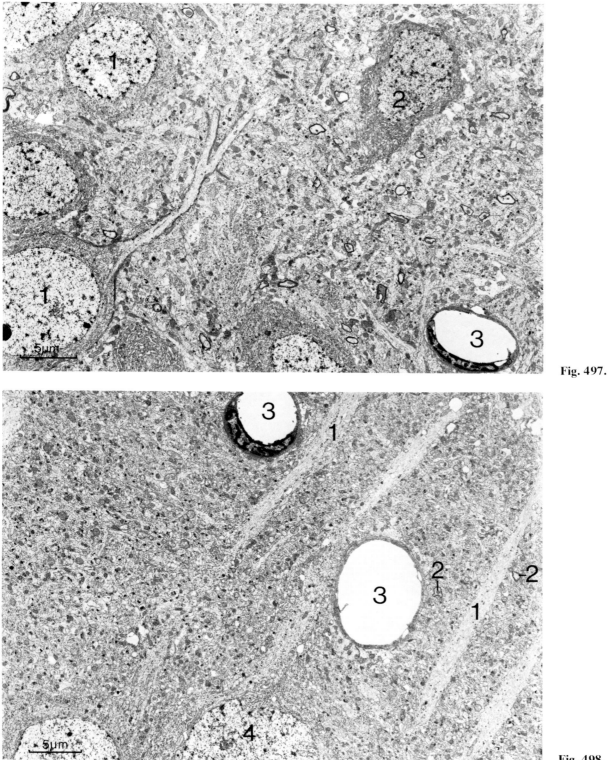

Fig. 497.

Fig. 498.

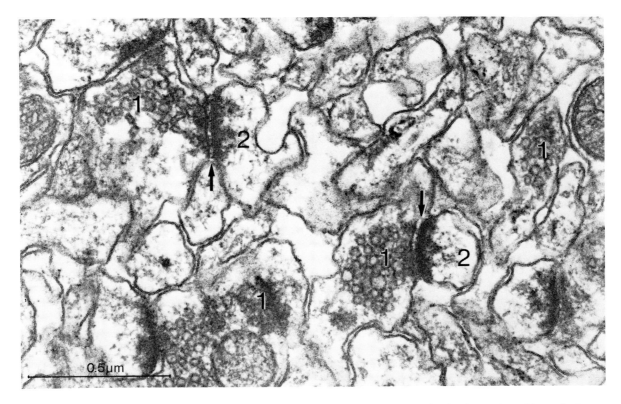

Fig. 499. A higher magnification illustrates that the neuropil resembles a puzzle in appearance in that it consists of irregularly outlined yet exactly fitting profiles whose definite classification can be difficult. Those cytoplasmic areas that are rich in vesicles (1) can be identified as presynaptic terminals of axons that make close (synaptic) contact (→) with adjoining dendrites. The synapses often show a marked density subjacent to the postsynaptic cell membrane (2). Magnification 74,000×.

◄

Fig. 497. Low-power electron micrograph of a rat cerebral cortex illustrating several nerve cells. These exhibit small round or ovoid perikarya containing a large vesicular nucleus (1). For a clearer demarcation against the surrounding neuropil, one dendrite is outlined in black. The darker staining and cone-shaped cell (2) is a slightly artificially damaged nerve cell rather than an oligodendrocyte. 3 = Capillary lumen. Magnification 3,000×.

Fig. 498. Low-power electron micrograph of a rat cerebral cortex illustrating its neuropil. This is defined as the sum of all kinds of glial and nerve cell processes interposed between the nerve cells, and it is traversed by several axons (1) in this micrograph. In the mass of its cytoplasmic profiles only the myelinated axons (2) can be identified, while the majority of cross-sectioned cell processes remain but poorly defined. 3 = Lumina of capillaries. Magnification 3,000×.

215

Tables

Table 1. Stains.

	Nuclei	Cytoplasm	Collagenous fibers	Elastic fibers
H.E. = hematoxylin and eosin	blue-violet	red	red	unstained or light pink
Mallory-azan = azocarmine and aniline blue modified after Mallory	red	light pink or bluish	blue	unstained (only when occurring in high concentrations as in elastic membranes and ligaments: red or reddish blue)
Elastica stain (resorcin-fuchsin or orcein) mostly combined with nuclear fast red (counterstain)	red	light pink	gray	blackish violet or dark brown
Van Gieson (iron-hematoxylin, picric acid and acid fuchsin)	black	yellow	red	no special staining (only in high concentrations as in elastic membranes and ligaments: yellow)
Trichrome stain after Masson-Goldner (iron-hematoxylin; Ponceau acid fuchsin; azophloxine/light green)	brownish black	orange-red	green	no special staining
E.H. = iron-hematoxylin after Heidenhain (particularly suitable for the staining of cell organelles, muscular cross-striations, etc.)	bluish black	—	—	light gray

Tables

Table 2.

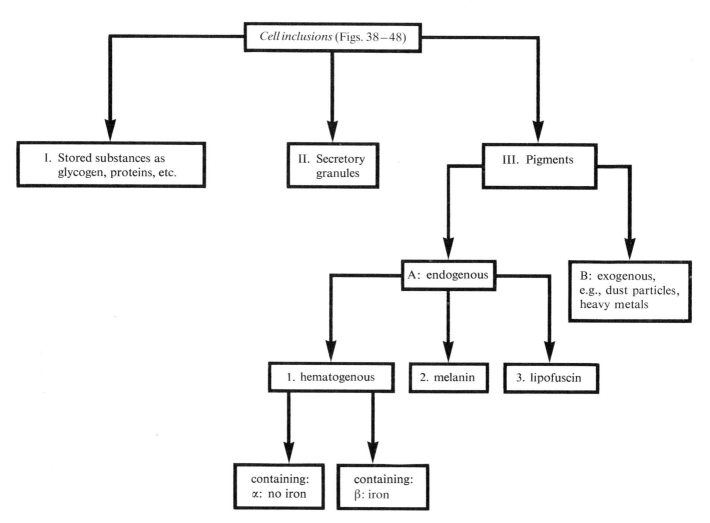

Cell inclusions (Figs. 38–48)

I. Stored substances as glycogen, proteins, etc.

II. Secretory granules

III. Pigments

A: endogenous

B: exogenous, e.g., dust particles, heavy metals

1. hematogenous

2. melanin

3. lipofuscin

containing: α: no iron

containing: β: iron

Table 3. Classification of epithelial tissues according to the shape of cells and their arrangement (after K. Zeiger).

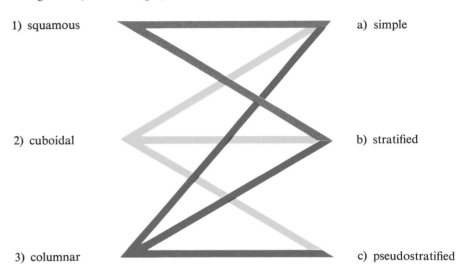

1) squamous   a) simple

2) cuboidal   b) stratified

3) columnar   c) pseudostratified

Table 4. Types of epithelial tissues and their locations.

1) squamous	a) *simple*	predominantly all meso- and endothelia
	b) *stratified*	α) cornified, e.g., skin
		β) noncornified, e.g., oral cavity, vagina, cornea, esophagus
2) cuboidal	a) *simple*	e.g., in excretory ducts, some kidney tubules, germinal epithelium of the ovary, etc.
	b) *stratified*	(infrequent) in some parts of excretory ducts
	c) *pseudostratified*	transitional epithelium
3) columnar	a) *simple*	α) with kinocilia: uterus, uterine tube
		β) without kinocilia: gastrointestinal tract, gall bladder
	b) *stratified*	(infrequent) conjunctival fornix, parts of the male and female urethra
	c) *pseudostratified*	α) without kinocilia: certain parts of glandular ducts (infrequent)
		β) with kinocilia: respiratory tract
		γ) with stereocilia: ductus epididymidis, ductus deferens

Table 5.   Principles for the classification of exocrine glands.

Morphological criteria		Examples
1) According to the number of secretory cells	unicellular glands multicellular glands	goblet cells salivary glands
2) According to the location of the secretory cells with regard to the epithelium	intraepithelial glands extraepithelial glands	goblet cells all large exocrine glands
3) According to the mechanisms of secretion	holocrine glands eccrine glands apocrine glands	sebaceous glands sweat glands prostate gland
4) According to the nature of their secretion	serous glands mucous glands mucoid glands	parotid gland goblet cells pyloric glands
5) According to the shape of their secretory units	tubular glands acinar glands alveolar glands mixed forms: tubulo-acinar tubulo-alveolar	crypts of Lieberkühn parotid gland apocrine sweat glands submandibular gland lactating mammary gland
6) According to the occurrence and the arrangement (e.g., branched or not) of a duct system	simple glands (each secretory portion empties separately on an epithelial surface) branched glands (several secretory units empty into an unbranched excretory duct) compound glands (secretory portions empty into an elaborate and branched duct system)	sweat glands pyloric glands all large salivary glands

Table 6. Family tree of the different types of the connective tissues (modified after K. Zeiger, 1948).

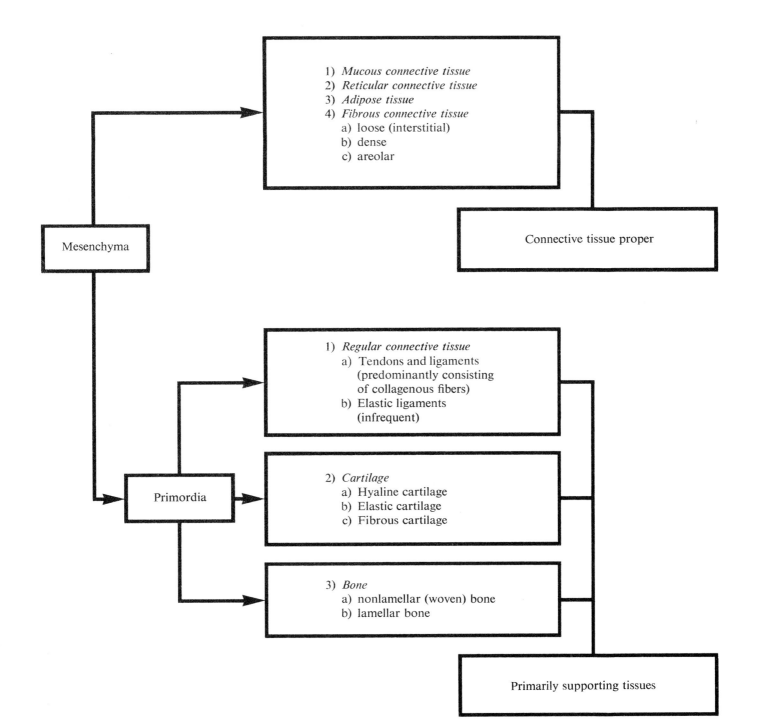

Table 7.   Various biological, light microscopical and staining properties of connective tissue fibers.

	Collagenous fibers	*Elastic fibers*	*Reticular fibers*
Arrangement	meshworks of varying texture	true networks or fenestrated membranes (e.g., inner elastic lamina)	delicate networks (regularly located at the interface between the interstitial connective tissue and the parenchymal cells of nearly every organ)
Microscopical appearance in fresh preparations	undulating course of longitudinally striated bundles of fibrils, poorly refractile	glassy, homogeneous (no fibrillar substructure), double contoured, highly refractile	not recognizable as such
Optical properties	highly anisotropic, hence showing uniaxial form and crystalline birefringence	isotropic in an unstretched state (hence no birefringence) but with increasing distension becoming anisotropic	slightly birefringent
Behavior in dilute acids	considerable swelling	—	moderate swelling
Behavior in dilute alkalis	dissolution	—	moderate dissolution
Mechanical properties	nonextensible	reversibly extensible to ca. 150% of their original length	moderately distensible
Staining properties Mallory-azan	blue	unstained, in high amounts as in elastic membranes: orange-red	blue
H.E.	red	unstained, in higher concentrations: light pink	—
Van Gieson	red	unstained, in higher concentrations: yellow	—

Table 8. Various types of "fibers".

1. *Connective tissue fibers* (collagenous, elastic and reticular): formed, noncellular components of the intercellular substance.

2. *Sharpey's fibers:* collagenous fibers passing from the periosteum into the bone.

3. *Nerve fibers:* a definite cellular process (= axon) that belongs to every ganglion cell.

4. *Lenticular fibers:* the extremely elongated nonnucleated apical portions of the equatorial epithelium.

5. *Tomes' fibers:* processes of the odontoblasts in the dentinal canals, hence also called dentinal fibers.

6. *Myocardial fibers:* either the macroscopically visible strands of the myocardium or a single strand consisting of closely applied myocardial cells one after the other.

7. *Smooth muscle fibers:* similar to those in the cardiac muscle, these fibers are composed of individual cells that are arranged into bundles mostly coursing in definite directions, e.g., the longitudinal and circular muscle layers of the intestine.

8. *Skeletal muscle fiber:* a tube-like multinucleated plasmodium representing the structural unit of the skeletal muscles.

9. *Glial fibers:* cytoplasmic processes of definite glial cells.

10. *Purkinje fibers:* Ultimate ramifications of the cardiac impulse conducting system.

Tables

Table 9.   Regularly recognizable and hence essential features for the differentiation of muscular tissues.

Type of tissue	Structural unit	Number of nuclei per structural unit	Location of the nuclei	Shape of the nuclei	Size of the structural units length	diameter
Skeletal muscle	fiber	several hundreds up to thousands	subsarcolemmal	elongated, flat	up to several cm	20–100 μm
Myocardium	cell	1–2	centrally with perinuclear cytoplasm free of myofibrils	plumpish round-ovoid	50–120 μm	10–20 μm
Smooth muscle	cell	1	centrally	elongated, rod-shaped or elliptical	40–200 μm (in a pregnant uterus up to 500 μm)	5–10 μm

Table 10.   Histological characteristics useful for identifying lymphatic organs.

	Tonsils	Lymph node	Thymus	Spleen
Epithelium	+	—	—	—
Connective tissue capsule	—	+	+	+
Organization into cortex and medulla	—	+	+	—
Marginal sinus	—	+	—	—
Hassall's corpuscles	—	—	+	—
Malpighian bodies	—	—	—	+

Table 11. Compilation of those regions that possess several surfaces mostly covered by different epithelia.

	Lips	Uvula	Epiglottis	Eyelids	Nostrils	Ear lobes	Portio vaginalis
Epithelium changes from:	Epidermis with hairs and various glands to the squamous, stratified and noncornified variety	Stratified squamous noncornified to a pseudo-stratified, columnar and ciliated epithelium	Stratified squamous noncornified to a pseudo-stratified, columnar and ciliated epithelium	Epidermis (without hair follicles) to a stratified, noncornified squamous epithelium	Epidermis with seba-ceous glands unconnected with hairs to an epidermal epithelium with hairs (vibrissae) and glands followed by a respiratory epithelium	Both surfaces are covered by the same epithelium: epidermis with typical cutaneous adnexes	Stratified non-cornified squa-mous epithe-lium (covering the outer sur-face) to a sim-ple columnar epithelium (lining the cervical canal)
Central tissue core predominantly consisting of:	Skeletal muscle (orbicularis oris muscle)	Skeletal muscle (uvular muscle)	Elastic cartilage	Skeletal muscle (orbicularis oculi muscle) and Meibo-mian glands	Hyaline cartilage	Elastic cartilage	Smooth muscle

Table 12. Differential diagnosis of the salivary glands, including the lacrimal gland.

Gland	Shape of secretory units	Duct system	Other characteristics
Parotid	Acinar ("serous")	Elaborate, conspicuous, large numbers of striated ducts (best criterion to distinguish parotid gland from exocrine pancreas)	Numerous fat cells together with sections of the arborizations of the facial nerve
Sub-mandibular	Tubulo-acinar (sero-mucous) with a prevailing serous (acinar) component	Well-developed	Serous demilunes capping the tubular (mucous) secretory units
Sublingual	Tubulo-acinar (sero-mucous) with a prevailing mucous (tubular) component	Intercalated and striated (salivary) ducts are sparse	Serous demilunes capping the tubular (mucous) secretory units
Lacrimal	Tubulo-alveolar, branched Serous (!) secretion	No intercalated and no striated ducts	In the connective tissue septa aggregations of free cells, preferentially plasma cells
Exocrine pancreas	Acinar ("serous")	No striated ducts, rest of the duct system much less developed than in parotid gland	Centro-acinar cells

Table 13.   Differential diagnosis of the consecutive segments of the digestive tract, including the gall bladder.

Part of the digestive tract	Folds	Villi	Crypts	Goblet cells	Special characteristics
Stomach, fundus	Sparse and coarse	—	—	—	Shallow gastric pits, deep fundic glands with chief and parietal cells
Stomach, pyloric part	Rare and coarse	—	—	—	Deep gastric pits, low pyloric glands, no chief and no parietal cell
Duodenum	Elaborate	+	+	+	Glands of Brunner within the submucosa, therefore also found within the folds
Jejunum	Numerous	+	+	+	
Ileum	Decreasing in number and lower	+	+	+	Aggregated lymphatic nodules, Peyer's patches, within the submucosa
Colon	Rare and coarse	—	+	+	Mucosa exclusively contains crypts but no villi
Vermiform appendix	—	—	+	+	Large focal lymphatic infiltrations within the submucosa and the mucosa
Gall bladder	Very delicate anastomosing folds	—	—	—	Missing layering of the muscular tunic (characteristic criterion for differential diagnosis!)

Table 14. Differential diagnosis of various "alveolar" glands including the fetal lung.

Gland	Lobular subdivisions	Duct system	Secretory units	Special characteristics
Prostate	Ill-defined	Almost nonexistent	Wide alveoli with a frill-like inner contour	Interstitial connective tissue crowded with smooth muscle (crucial criterion for diagnosis)
Lactating mammary	Very distinct	Excretory ducts are regularly seen in the inter-lobular septa (crucial criterion for diagnosis)	Varying in size, lipid vacuoles within the secretory cells	
Thyroid	Distinct	None	Follicles are the largest of all alveolar secretory units, size and lumen vary	"Secretory units" (= follicles) filled with deeply staining material (colloid)
Fetal lung	Distinct, conspicuous cellular connective tissue	Always distinct	Often appearing as branching epithelial tubules	Close to the "duct" system (primordia of bronchi) hyaline cartilage can be found (crucial criterion for diagnosis)

Table 15.   Differential diagnosis of hollow organs showing a stellate lumen in cross section.

Organ	Epithelium	Muscular tunic	Special characteristics
Esophagus	Squamous, stratified non-keratinizing	Very prominent, distinctly divided into inner circular and outer longitudinal layer	Prominent muscularis mucosae, scattered small glands within the submucosa
Ureter	Transitional	More loosely arranged, subdivided into a prominent intermediate circular layer upon which inner and outer bundles of longitudinal muscles are loosely attached	
Urethra	Columnar, bi- to four-layered, pseudostratified or stratified	No distinct layering, very loosely arranged muscular network	Prominent venous plexus within the lamina propria
Ductus deferens	Columnar, pseudostratified with stereocilia	Very prominent, distinctly arranged into inner longitudinal; intermediate circular and outer longitudinal layer	Often sectioned together with the entire spermatic cord
Uterine tube	Columnar, simple with kinocilia	Relatively thick, no distinct layering, preferentially circular	Slender, richly arborized mucosal folds

Subject Index

Index

Index